SEMI-CONDUCTOR MONOGRAPHS

OPTICAL PROPERTIES
OF SEMI-CONDUCTORS

T. S. MOSS, M.A., Ph.D.

Royal Aircraft Establishment
Farnborough, Hants

NEW YORK
ACADEMIC PRESS INC., PUBLISHERS

LONDON
BUTTERWORTHS SCIENTIFIC PUBLICATIONS
1959

BUTTERWORTHS PUBLICATIONS LTD
88 KINGSWAY, LONDON, W.C.2

U.S.A. Edition published by
ACADEMIC PRESS INC., PUBLISHERS
111 FIFTH AVENUE
NEW YORK 3, NEW YORK

Printed in Northern Ireland at The Universities Press, Belfast

PREFACE

DURING the past few years there has been a great expansion in the amount of interest and effort—both academic and commercial—devoted to semi-conductors, with a corresponding increase in the number of papers published on various semi-conductor topics. The volume of this literature is now so great that the need for books which review the field is quite apparent.

The few books which have been published on semi-conductors so far have provided a rather general coverage, and it is felt that the subject has now expanded to a degree where individual books dealing with only a fraction of the total field are merited. The present work falls in this category.

The first forty per cent of the book gives a theoretical treatment of the optical properties of conducting media, absorption and emission processes, inter-relation of the optical constants, magneto-optic effects and all internal photo-effects. The remaining sixty per cent of the book is devoted to an account of the properties of many of the most interesting semi-conductors.

In such a rapidly moving field as this, with such intensive and extensive research and development work in progress, it is of paramount importance that a book should be up to date, and particular efforts have been made to achieve this end. Several of the most important chapters were first drafted in 1958, all chapters have been revised this year, and many of the important results of the International Semi-conductor Conference held at Rochester in August 1958 are included.

The book is primarily intended for the experimental physicist who is engaged in active academic research or commercial development work on semi-conductors and for the honours student who wishes to increase his specialized knowledge, perhaps because he may soon become an active worker in this field himself.

I wish to take this opportunity of acknowledging the assistance of my wife in the production of the book, particularly for preparing the manuscript, checking the proofs and forming the indices.

TREVOR SIMPSON MOSS

Farnborough, Hants,
September, 1958

CONTENTS

CHAPTER 1

THEORY OF OPTICAL PROPERTIES OF CONDUCTING MEDIA

1.1 WAVE PROPAGATION

THE theory of the propagation of electro-magnetic waves in conducting materials is based on Maxwell's field equations, which may be written:

$$\operatorname{curl} \tilde{E} = -\mu\mu_0 \frac{\partial \tilde{H}}{\partial t} \tag{1.1a}$$

$$\operatorname{curl} \tilde{H} = \sigma\tilde{E} + \varepsilon\varepsilon_0 \frac{\partial \tilde{E}}{\partial t} \tag{1.1b}$$

$$\operatorname{div} \tilde{H} = 0 \tag{1.1c}$$

$$\operatorname{div} \tilde{E} = 0 \tag{1.1d}$$

where ε_0 and μ_0 are the dielectric constant and permeability of free space and μ and ε are the specific permeability and dielectric constant of the medium and the other symbols have their usual meaning. Equation (1.1d) is taken as zero since in a conducting medium there can be no permanent charge density.

Hence

$$\operatorname{curl} \operatorname{curl} \tilde{E} = -\mu\mu_0 \left(\sigma \frac{\partial \tilde{E}}{\partial t} + \varepsilon\varepsilon_0 \frac{\partial^2 \tilde{E}}{\partial t^2} \right)$$

But

$$\operatorname{curl} \operatorname{curl} \tilde{E} = \operatorname{grad} \operatorname{div} \tilde{E} - \nabla^2 \tilde{E}$$

Therefore

$$\nabla^2 \tilde{E} - \sigma\mu\mu_0 \frac{\partial \tilde{E}}{\partial t} - \mu\mu_0\varepsilon\varepsilon_0 \frac{\partial^2 \tilde{E}}{\partial t^2} = 0 \tag{1.2}$$

A similar equation is obtained for $\tilde{H}$.

Write as a solution for one of the components of $\tilde{E}$ or $\tilde{H}$

$$U_x = U_0 e^{i\omega(t-x/v)} \tag{1.3}$$

This solution satisfies the equation provided

$$1/v^2 = \mu\varepsilon\mu_0\varepsilon_0 - i\sigma\mu_0\mu/\omega \tag{1.4}$$

Expression (1.3) represents a wave travelling in the x direction with

1

velocity v. Now $v = c/N$ where c is the velocity of light *in vacuo* and N is the refractive index of the medium. Hence

$$N^2 = c^2(\mu\varepsilon - i\sigma\mu/\omega\varepsilon_0)\mu_0\varepsilon_0 \qquad (1.5)$$

For free space where $N = 1$, $\varepsilon = 1$, $\mu = 1$ and $\sigma = 0$,

$$1 = c^2\varepsilon_0\mu_0 \qquad \text{or} \qquad c = (\varepsilon_0\mu_0)^{-1/2}$$

Hence

$$N^2 = \mu\varepsilon - i\sigma\mu/\omega\varepsilon_0 \qquad (1.6)$$

Clearly for non-vanishing conductivity the refractive index is complex and may be expressed as

$$N = n - ik \qquad (1.7)$$

giving

$$U_x = U_0 e^{i\omega t} e^{-i\omega n x/c} e^{-\omega k x/c}$$

This expression is now seen to represent a wave of frequency $\omega/2\pi$ travelling with velocity c/n and suffering attenuation or absorption.

From equations (1.6) and (1.7)

$$n^2 - k^2 = \mu\varepsilon \qquad 2nk = \sigma\mu/\omega\varepsilon_0$$

For all cases of practical interest in optical properties we can take $\mu = 1$ and hence

$$n^2 - k^2 = \varepsilon \qquad 2nk = \sigma/\omega\varepsilon_0 \qquad (1.8)$$

giving

$$\begin{aligned} 2n^2 &= \varepsilon[1 + (1 + \sigma^2/\omega^2\varepsilon^2\varepsilon_0^2)^{1/2}] \\ 2k^2 &= \varepsilon[1 - (1 + \sigma^2/\omega^2\varepsilon^2\varepsilon_0^2)^{1/2}] \end{aligned} \qquad (1.9)$$

It should be noted that σ is the conductivity at the optical frequency concerned, and is not generally equal to the d.c. or low frequency conductivity σ_0.

Clearly as $\sigma \to 0$, $k \to 0$ and $n^2 \to \varepsilon$. It has been verified experimentally that the latter relation holds provided the measurement of n and ε are made at the same frequency, or, if made at different frequencies, as long as there are no absorption bands at any intermediate frequencies.

The absorption coefficient for the medium, K, is defined by the condition that the energy in the wave falls by $e:1$ in a distance $1/K$. As the energy flow is given by the Poynting vector it is thus proportional to the product of the amplitudes of the electric and magnetic vectors. As both these contain the term $e^{-\omega k x/c}$ the attenuation is $e^{-2\omega k x/c}$ and the absorption coefficient

$$K = 2\omega k/c = 4\pi k/\lambda \qquad (1.10)$$

where λ is the wavelength in free space. Measurements of transmission through samples of the material of different thicknesses may thus be used to determine K and k directly.

As equation (1.8) shows that in general $n^2 \neq \varepsilon$, it should be pointed out that the velocity of the wave is given by c/n not $c/\varepsilon^{1/2}$. Hence measurements of properties determined by the velocity of the wave alone will give a direct value for n.

In general experiments of the types described above to determine n and k separately can only be performed on material where the absorption is not intense—i.e. where specimens can be prepared of thickness only a few times $1/K$. For highly absorbing materials where the optical properties more nearly resemble metals than dielectrics it is necessary to rely on the analysis of reflection measurements using polarized radiation which usually yields two simultaneous equations involving both n and k.

1.2 Behaviour at a conducting surface

The problem of the behaviour of a plane wave at the intersection between two media becomes considerably more difficult when one of the media is conducting. Snell's and Fresnel's laws still hold, but the interpretation is complicated by the fact that the angle of refraction is imaginary, and in the refracted wave the planes of constant phase and of constant amplitude no longer coincide.

Consider a plane wave travelling in a dielectric medium of refractive index N incident on the interface of a conducting medium where the complex refractive index is N''. The permeabilities μ_1 and μ_2 are taken as unity. Let the boundary of the two media be at $z = 0$, and with the incident wave in the xz plane. Resolve the electric intensity E_0 in this wave into components E_n normal to, and E_p parallel to the plane of incidence. E_n thus lies in the y direction whilst E_p has components $E_p \cos \phi$ and $-E_p \sin \phi$ in the x and z directions respectively, where ϕ is the angle between the incident wave and the surface normal. The incident wave may be written:

$$E_y = E_n \exp i\omega[t - N(x \sin \phi + z \cos \phi)/c]$$
$$E_x \sec \phi = -E_z \operatorname{cosec} \phi = E_p \exp i\omega[t - N(x \sin \phi + z \cos \phi)/c]$$

$$(1.11)$$

The magnetic components may be found from equation (1.1a) which gives

$$\tilde{H} = -\frac{\operatorname{curl} \tilde{E}}{i\omega \mu_0} \qquad (1.12)$$

3

Similarly for the reflected wave:

$$\left.\begin{aligned}
E'_y &= E'_n \exp i\omega[t - N(x \sin \phi' + z \cos \phi')/c] \\
E'_x \sec \phi' &= -E'_z \operatorname{cosec} \phi' \\
&= E'_p \exp i\omega[t - N(x \sin \phi' + z \cos \phi')/c]
\end{aligned}\right\} \quad (1.13)$$

and for the refracted wave

$$\left.\begin{aligned}
E''_y &= E''_n \exp i\omega[t - N''(x \sin \phi'' + z \cos \phi'')/c] \\
E''_x \sec \phi'' &= -E''_z \operatorname{cosec} \phi'' \\
&= E''_p \exp i\omega[t - N''(x \sin \phi'' + z \cos \phi'')/c]
\end{aligned}\right\} \quad (1.14)$$

It is clear that at the boundary where $z = 0$ all the components of E and H must vary with the same function of x. Hence

$$N \sin \phi = N \sin \phi' = N'' \sin \phi'' \qquad (1.15)$$

From (1.15) Snell's laws follow directly*, namely, (i) reflection, $\phi = \pi - \phi'$, i.e. the angle of incidence equals the angle of reflection, (ii) refraction, $N \sin \phi = N'' \sin \phi''$.

Fresnel's laws concerning the amplitudes of the reflected and refracted waves are given by the boundary conditions that the tangential components of the electric and magnetic intensities do not change on passing through the plane $z = 0$. For the electric intensity

$$\left.\begin{aligned}
E_p \cos \phi + E'_p \cos \phi' = E''_p \cos \phi'' &= (E_p - E'_p) \cos \phi \\
E_n + E'_n &= E''_n
\end{aligned}\right\} \quad (1.16)$$

Similarly the condition for the magnetic intensity to be continuous through $z = 0$ is found to give

$$\left.\begin{aligned}
(E_n - E'_n) N \cos \phi &= E''_n N'' \cos \phi'' \\
(E_p + E'_p) N &= E''_p N''
\end{aligned}\right\} \quad (1.17)$$

Solving these four equations gives for the reflected wave amplitudes

$$E'_p = E_p \frac{N'' \cos \phi - N \cos \phi''}{N'' \cos \phi + N \cos \phi''} = E_p \frac{\tan (\phi - \phi'')}{\tan (\phi + \phi'')} \quad (1.18a)$$

$$E'_n = E_n \frac{N \cos \phi - N'' \cos \phi''}{N \cos \phi + N'' \cos \phi''} = E_n \frac{\sin (\phi'' - \phi)}{\sin (\phi + \phi'')} \quad (1.18b)$$

* For a more detailed discussion see Stratton (1941).

and for the refracted wave amplitudes

$$E''_p = E_p \frac{2N \cos \phi}{N \cos \phi'' + N'' \cos \phi} = E_p \frac{2 \sin \phi'' \cos \phi}{\sin (\phi + \phi'') \cos (\phi - \phi'')}$$

$$(1.19a)$$

$$E''_n = E_n \frac{2N \cos \phi}{N'' \cos \phi'' + N \cos \phi} = E_n \frac{2 \sin \phi'' \cos \phi}{\sin (\phi + \phi'')} \quad (1.19b)$$

These are Fresnel's equations. Corresponding equations are obtained for the magnetic vector. The refraction angle ϕ'' may be eliminated by use of equation (1.15).

When the wave is normal to the surface the reflected amplitude reduces to

$$E' = E'_p = E'_n = E(N'' - N)/(N'' + N) \qquad (1.20)$$

The energy flow is given by the real part of the complex Poynting vector so that in the initial wave,

$$S_0 = \tfrac{1}{2} E_0 \times H^*_0 = \tfrac{1}{2} N E_0{}^2$$

in the direction of propagation of the incident wave. Similarly for reflection and transmission

$$S_r = \tfrac{1}{2} N (E')^2 \qquad S_t = \tfrac{1}{2} N'' (E'')^2$$

Now the reflection and transmission coefficients are defined as the ratios of the energy flows normal to the surface, so that

$$T = (E''/E)^2 \cos \phi''/\cos \phi \qquad R = (E'/E)^2$$

Hence when the electric vector is normal to the plane of incidence

$$R_n = \frac{\sin^2 (\phi'' - \phi)}{\sin^2 (\phi + \phi'')} \qquad T_n = \frac{\sin 2\phi \sin 2\phi''}{\sin^2 (\phi + \phi'')} \qquad (1.21)$$

and when the electric vector is parallel to the plane of incidence

$$R_p = \frac{\tan^2 (\phi - \phi'')}{\tan^2 (\phi + \phi'')} \qquad T_p = \frac{\sin 2\phi \sin 2\phi''}{\sin^2 (\phi + \phi'') \cos^2 (\phi - \phi'')} \qquad (1.22)$$

In all cases $R + T = 1$, as may be seen from the fact that the normal component of energy flow must be continuous at the interface.

In the simple case of non-absorbing media these expressions for R and T are real and may be used as they stand. For normal incidence on an interface between media of indices n_1 and n_2 the expressions reduce to:

$$\left. \begin{array}{l} R = R_p = R_n = (n_2 - n_1)^2/(n_2 + n_1)^2 \\ T = T_n = T_p = 4n_2 n_1/(n_2 + n_1)^2 \end{array} \right\} \qquad (1.23)$$

One other particular case is of interest, namely when $\phi + \phi'' = 90°$ for then $R_p = 0$. This condition is given by $\tan \phi = N''/N$ and for non-absorbing media ϕ is a real angle, the Brewster angle. Thus for a pure dielectric the refractive index may be determined directly by allowing a beam of plane polarized light (with the electric vector parallel to the plane of incidence) to fall on to the surface of the dielectric and adjusting the angle of incidence until there is no

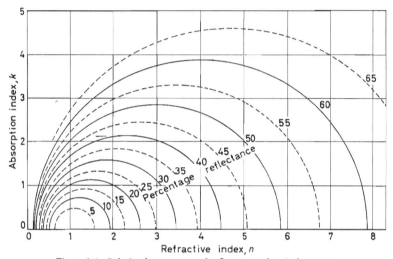

Figure 1.1. Relation between normal reflectance and optical constants

reflection. As will be seen later, when the second medium is conducting, R_p never vanishes, although when the conductivity is low there is still a pronounced minimum for angles of incidence near $\tan^{-1} (N''/N)$. It may be noted that as $\phi + \phi''$ passes through the 90° value there will be an abrupt change of phase of 180° in the reflected wave.

1.2.1 Reflection normal to conducting surface

Consider the case where the first medium is air with $N = 1$ and for the second medium $N'' = n - ik$. Equation (1.23) is valid and gives

$$R = (n - ik - 1)^2/(n - ik + 1)^2$$

which becomes, on rationalizing,

$$R = [(n - 1)^2 + k^2]/[(n + 1)^2 + k^2] \tag{1.24}$$

For a material of low absorption where k is small, R is little more

6

than for the pure dielectric, but as k becomes large the material behaves like a metal and the reflectivity approaches unity.

Equation (1.24) for the reflectivity at normal incidence may be rewritten

$$k^2 + \left(n - \frac{1+R}{1-R}\right)^2 = \frac{4R}{(1-R)^2}$$

The equation is seen to represent a family of circles with centres at $n = (1+R)/(1-R)$ of radii $2R^{1/2}/(1-R)$. Such a set of curves are shown in *Figure 1.1*.

1.3 REFLECTIVITY AT AIR–CONDUCTOR INTERFACE

The procedure to be followed is the same as for the case of normal incidence (except that it is considerably more tedious) namely to insert $N = 1$ and $N'' = n - ik$ in the appropriate equations (1.21) and (1.22) and rationalize. The solutions show that in general the reflected wave is no longer plane polarized but is elliptically polarized. Put

$$N'' \cos \phi'' = [(N'')^2 - \sin^2 \phi]^{1/2} = \alpha + i\beta \qquad (1.25)$$

where

$$\alpha^2 - \beta^2 = n^2 - k^2 - \sin^2 \phi' \qquad \alpha\beta = -nk \qquad (1.26)$$

Then from (1.15)

$$R_n = \left|\frac{\sin(\phi'' - \phi)}{\sin(\phi'' + \phi)}\right|^2 = \left|\frac{\cos \phi - N'' \cos \phi''}{\cos \phi + N'' \cos \phi''}\right|^2$$

$$= \frac{(\alpha - \cos \phi)^2 + \beta^2}{(\alpha + \cos \phi)^2 + \beta^2} \qquad (1.27)$$

From (1.16),

$$R_p = \left|\frac{\sin(\phi'' - \phi)}{\sin(\phi'' + \phi)} \frac{\cos(\phi'' + \phi)}{\cos(\phi'' - \phi)}\right|^2$$

$$= R_n \left[\frac{(\alpha - \sin \phi \tan \phi)^2 + \beta^2}{(\alpha + \sin \phi \tan \phi)^2 + \beta^2}\right] \qquad (1.28)$$

For semi-conductors, in the majority of cases of interest, n is several times unity while k is about unity or less. Hence $n^2 \gg k^2$ and $n^2 \gg \sin^2 \phi$. It follows that $\alpha \sim n$, $\beta \sim k$. It may thus be seen that R_n is a slowly varying function of ϕ, rising from a typical value of ~30 per cent at normal incidence to 100 per cent at glancing incidence (see *Figure 1.2*). R_p has the same value as R_n at $\phi = 0$, but passes through a minimum near $\alpha = \sin \phi \tan \phi$, i.e. when $\tan \phi \sim n$, and then rises rapidly to 100 per cent at $\phi = 90°$. For

7

values of n between 3 and 6 this angle of minimum reflectivity lies between 71° and 81°.

If ϕ is measured there are only two unknowns (α and β) in equations (1.27) and (1.28), and so their values (and hence n and k) may be determined by any two measurements. Equation (1.27) is

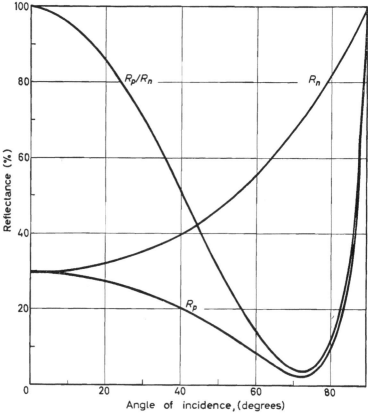

Figure 1.2. Theoretical reflectance curves for $n = 3$, $k = 1$

the simpler to analyse, but as it only varies slowly with n and k it does not possess good inherent accuracy. R_p has the virtue that it varies rapidly with n and k in the neighbourhood of $\phi = \tan^{-1} n$ and measurements near this angle are potentially capable of giving good accuracy of n and k. By measuring the ratio R_p/R_n the computation is considerably simplified, and furthermore the accuracy of the experiment is improved as this ratio varies more rapidly with angle of incidence than R_p alone (see *Figure 1.2*) and the equipment is no longer called upon to measure absolute reflectivities*. This method has been used by Avery (1952, 1953) who solved the tedious computational problem by the use of sets of curves of values of R_p/R_n computed for feasible values of n and k at two or three specific angles of incidence. These angles should be chosen so that one is just below and one just above the angle of minimum reflectivity—for example 65° and 83°—with perhaps a third intermediate angle which must inevitably be fairly near the critical angle. Measurements at two angles give only one estimate of n and k, measurements at three angles may be averaged for n and k and the consistency of the three values of n and k obtained is a useful indication of the accuracy of the experiment.

If k is small, R_p/R_n has a sharp minimum at the pseudo-Brewster angle given by $\sin^2 \phi \tan^2 \phi = \alpha^2 + \beta^2$, which approximates closely to

$$\tan^2 \phi \doteqdot n^2 + k^2 \tag{1.29}$$

also, for high index materials,

$$(R_p/R_n)_{\min} = \beta^2/4\alpha^2 \doteqdot k^2/(4n^2 - 4) \tag{1.29a}$$

For the absolute reflectivity with the electric vector in the plane of incidence, Miller and Johnson (1954) give a minimum value

$$(R_p)_{\min} = k^2(1 - 2/n^2)/4n^2 \tag{1.29b}$$

As is clear from *Figure 1.2* this expression must be smaller than that given by (1.29a).

Measurements of (1.29b) or preferably (1.29a), are probably as good as any polarized reflection experiments for the determination of low values of k. Even so, if $(R_p/R_n)_{\min} = 1$ per cent is taken as a reasonable limit of accurate measurement, this corresponds to $k = 0.6$ for $n = 3$. These figures clearly emphasize the fact that methods of measurement based on the analysis of reflected polarized light cannot be used for the accurate determination of values of k much less than unity.

Another limitation to the method of reflection measurement is imposed by the state of the surface of the material. According to Houston (1938), in a reflection measurement on a dielectric, a surface layer only $\sim\lambda/50$ thick is involved (see Appendix A). The presence of a surface film only $\lambda/200$ thick has been found to raise the

---

* In principle the optical constants may be obtained by measurement of the phase differences between incident and reflected beams with a suitable analyser and the ratio of R_p/R_n (Ditchburn, 1955).

reflectivity of glass at the Brewster angle from its ideal value of zero to $R_p \sim 10^{-4}$ corresponding to 1 per cent ellipticity in the reflected light (Wood, 1933). It is thus of great importance to ensure that the surface layers of the specimen are characteristic of the bulk material, and are not contaminated by grease or oxidation films or suffering from strains and distortions as a result of optical polishing.

1.4 REFRACTION AT A CONDUCTING SURFACE

As stated above, the behaviour of the refracted ray is still described by the generalized form of Snell's law

$$N \sin \phi = N'' \sin \phi''$$

or for the air/conductor interface treated above

$$\sin \phi = (n - ik) \sin \phi'' \qquad (1.30)$$

where $\sin \phi''$ is complex. The direction of propagation in the second medium is normal to the planes of constant phase in the medium. If this direction lies at an angle ψ to the surface normal then one can write a modified form of Snell's law as

$$n' = \frac{\sin \phi}{\sin \psi} = (\sin^2 \phi + \xi^2)^{1/2} \qquad (1.31)$$

This quantity n' is a real index of refraction but it now depends on the angle of incidence and is therefore no longer characteristic of the material. The parameter ξ is given by the equation,

$$2\xi^2 = n^2 - k^2 - \sin^2 \phi + [4n^2k^2 + (n^2 - k^2 - \sin^2 \phi)^2]^{1/2} \quad (1.32)$$

With this unwieldy form of Snell's law it is no longer convenient to try to measure n by refraction experiments—using prisms for example—when k is high, even if the prism can be made sufficiently thin to give adequate transmission. When the absorption is not too great so that $k^2 \ll n^2$

$$\sin \phi \doteqdot n\left(1 + \frac{k^2}{2n^4}\right) \sin \psi \qquad (1.33)$$

Thus for $n = 3$ as occurs for many semi-conductors and $k = 1$, $\sin \psi$ differs by only $\frac{1}{2}$ per cent from the value it would have if k were zero. Little error would therefore result in refraction measurements of n provided k were less than unity. In practice, if $k = 1$ at wavelengths in the near infra-red, then the absorption coefficient $K > 10^4$ cm^{-1}. It thus appears that for any prism of macroscopic size, the simple form of Snell's law will be applicable with fair accuracy to any wave which has measurable transmission.

It should be noted that for the high refractive indices met in semi-conductors the prisms used must have a small angle, otherwise total internal reflection will prevent the emergence of the radiation. For the limiting case of 90° angles of incidence and emergence the prism angle is $2 \sin^{-1} (1/n)$ and for convenient measuring conditions the angle should not exceed $2 \sin^{-1} (0{\cdot}8/n)$. For $n \sim 5$, as occurs for tellurium, this angle is about 20°.

1.5 OPTICAL INTERFERENCE PHENOMENA, AND DETERMINATION OF n AND k

When a ray of light is incident on a thin parallel sheet of transparent material, multiple reflections take place inside the sheet, and at selected wavelengths constructive or destructive interference will occur in the reflected and transmitted beams. If a spectral measurement of transmission is carried out, a series of fringes will be obtained. Such a measurement is potentially the most accurate way of determining the refractive index. Also, as stressed in section 1.1, the velocity of the wave is independent of the value of k, although there are phase changes at reflections if k is not small.

Consider the behaviour of a ray of unit amplitude travelling through a medium of refractive index n_0 on to a specimen of index n_1 and thickness d mounted on a base of index n_2. The only phase change on reflection will be that of the directly reflected ray, which is thus designated $-r_1$. Internal reflection at the top and bottom surfaces respectively are given by r_1 and r_2, the transmission into the top surface by t_1, and the transmission out of the top and bottom surfaces by t' and t_2 respectively.

The phase change on one traversal of the film is

$$\delta = (2\pi/\lambda)n_1 d \cos \phi \qquad (1.34)$$

Hence the amplitude of the reflected beam relative to the incident beam of unit amplitude will be given by

$$\begin{aligned} r &= -r_1 + t_1 t' r_2 e^{-2i\delta} + t_1 t' r_1 r_2^2 e^{-4i\delta} + t_1 t' r_1^2 r_2^3 e^{-6i\delta} \\ &= -r_1 + tt' r_2 e^{-2i\delta}(1 - r_1 r_2 e^{-2i\delta})^{-1} \end{aligned} \qquad (1.35)$$

Similarly for transmission

$$\begin{aligned} t &= t_1 t_2 e^{-i\delta} + t_1 t_2 r_1 r_2 e^{-3i\delta} + t_1 t_2 r_1^2 r_2^2 e^{-5i\delta} \\ &= t_1 t_2 e^{-i\delta}(1 - r_1 r_2 e^{-2i\delta})^{-1} \end{aligned} \qquad (1.36)$$

The above formulae simplify for an unmounted film where

11

$n_2 = n_0$, $t_2 = t'$ and $r_1 = r_2$. Noting that $t_1 t' = 1 - r_1^2$ we obtain

$$r = -r_1 + r_1(1 - r_1^2)e^{-2i\delta}(1 - r_1^2 e^{-2i\delta})^{-1} \qquad (1.37)$$

$$t = (1 - r_1^2)e^{-i\delta}(1 - r_1^2 e^{-2i\delta})^{-1} \qquad (1.38)$$

Maximum transmission occurs when the phase change is an integral number of half wavelengths per traversal, i.e. $\delta = N\pi$ and $e^{-2i\delta} = 1$. Hence $t_{max} = e^{-i\delta}$ so that the maximum transmitted amplitude is of course unity, with minimum reflection correspondingly zero. Maximum reflectivity occurs when $2n_1 d \cos \phi = (N + \frac{1}{2})\lambda$, i.e. for $\delta = (N + \frac{1}{2})\lambda$ and $e^{-2i\delta} = -1$.

Hence

$$r_{max} = -r_1 - \frac{(1 - r_1^2)r_1}{1 + r_1^2} = \frac{-2r_1}{1 + r_1^2} \qquad (1.39)$$

and

$$t_{min} = \frac{1 - r_1^2}{1 + r_1^2} e^{-i\delta} \qquad (1.40)$$

The energy transmission is given by the product of t and its complex conjugate, and hence

$$T = tt^* = \left[\frac{(1 - r_1^2)e^{-i\delta}}{1 - r_1^2 e^{-2i\delta}}\right]\left[\frac{(1 - r_1^2)e^{i\delta}}{1 - r_1^2 e^{2i\delta}}\right]$$

$$= \frac{(1 - r_1^2)^2}{1 - 2r_1^2 \cos 2\delta + r_1^4} \qquad (1.41)$$

Hence $T_{max} = 1$ and $T_{min} = (1 - r_1^2)^2/(1 + r_1^2)^2$. Similarly $R_{max} = 4r_1^2(1 + r_1^2)^{-2}$. It may be noted that these two latter results could be obtained directly by squaring equations (1.39) and (1.40).

The visibility of fringes depends on the contrast, i.e. T_{max}/T_{min} for transmission fringes. For a typical semi-conductor with an index of $n_1 = 3$, $r_1 = (n_1 - 1)/(n_1 + 1) = \frac{1}{2}$, and $T_{max}/T_{min} \doteqdot 3$. For most semi-conductors therefore, the fringes are well defined in either transmission or reflection, and it is not usually necessary to plate the surfaces to increase the reflectivity.

The contrast between fringe maxima and minima is reduced if there is absorption in the film. Suppose the transmission is α for each traversal of the film. Then

$$\left.\begin{array}{l} t_{max} = \alpha(1 - r_1^2)(1 - \alpha^2 r_1^2)^{-1} \\[4pt] t_{min} = \alpha(1 - r_1^2)(1 + \alpha^2 r_1^2)^{-1} \\[4pt] r_{max} = -r_1(1 + \alpha^2)(1 + \alpha^2 r_1^2)^{-1} \\[4pt] r_{min} = -r_1(1 - \alpha^2)(1 - \alpha^2 r_1^2)^{-1} \end{array}\right\} \qquad (1.42)$$

In finding the refractive index from interference fringes, there is the difficulty that in general the order of the fringe (i.e. the value of N) is not known. The usual method of overcoming this difficulty is to measure the wavelengths of two adjacent transmission maxima for example and apply the equations

$$2nd = N\lambda_1 \tag{1.43a}$$

$$2nd = (N + 1)\lambda_2 \tag{1.43b}$$

Eliminating N, $2nd = (\lambda_2^{-1} - \lambda_1^{-1})^{-1}$.

It is not generally realized that this procedure is valid only if n is known to have no linear dependence on wavelength over the range considered. If n is varying and may be expressed as $n = n' + \beta\lambda$, then

$$2(n' + \beta\lambda_1)d = N\lambda_1 \qquad 2(n' + \beta\lambda_2)d = (N + 1)\lambda_2 \tag{1.44}$$

giving $2n'd(\lambda_2^{-1} - \lambda_1^{-1}) = 1$. Thus β is eliminated and may thus *have any value*. Furthermore, the value of index thus found is not the value at the wavelength concerned, but is n', the value extrapolated to $\lambda = 0$. This effect has been shown to be important in InSb. (See section 16.8 and Moss *et al.*, 1957.)

For the classical dispersion curve for a single oscillator as given by equation (2.8) there is of course, no term linear in λ (or ω), and if experimental data can be taken over a sufficient range of wavelengths to establish the form of the dispersion curve the lack of the linear term can be verified. To avoid uncertainty it is advisable for measurements to be made down to such low fringe orders that the value of N can be established unambiguously, and the refractive index then calculated at each wavelength from equation (1.43a).

The complete expression for the transmission of an absorbing layer may be obtained from equations (1.35) and (1.36) by replacing δ by $\delta + i\beta$ where

$$\beta = 2\pi kd/\lambda = \tfrac{1}{2}Kd \tag{1.45}$$

since there is an attenuation in amplitude of $\exp \tfrac{1}{2}Kd$ per traversal. Hence the transmission is given by

$$\left(tt^* = \frac{(1 - R)^2(1 + k^2/n^2)}{(e^\beta - Re^{-\beta})^2 + 4R\sin^2(\delta + \psi)} \right) \tag{1.46}$$

where the phase angle is given by equation (1.34), R is given by equation (1.24) and

$$\tan \psi = 2k/(n^2 + k^2 - 1) \tag{1.47}$$

13

If specimen imperfections or the spectral bandwidths used experimentally are such that the fringes are not resolved, then the transmission observed is the average of equation (1.46), which may readily be shown to be

$$T_{av} = (1 - R)^2(1 + k^2/n^2)/(\exp Kd - R^2 \exp -Kd) \quad (1.48)$$

In any practical transmission experiments on semi-conductors, $k^2 \ll n^2$. Also, it is usually true that

$$\exp 2Kd \gg R^2$$

so that the average transmission is simply

$$T_{av} = (1 - R)^2 \exp (-Kd) \qquad (1.49)$$

In analysing transmission measurements to find K, the $1 - R$ term may be allowed for by calculating R if the refractive index is known at the appropriate wavelength, or may be eliminated by measuring specimens of more than one thickness. Alternatively the reflectivity may be measured directly on a thick (i.e. totally absorbing) specimen, which is then ground down until it is thin enough for transmission measurements. This latter method has been used by Spitzer and Fan (1957b).

Detailed treatments of the optical behaviour of multi-layer 'sandwiches' are given by Heavens (1955) and Heavens and Smith (1957).

CHAPTER 2

DISPERSION THEORY

IN the theoretical treatment of dispersion it is convenient to consider separately the contributions from the bound and the free electrons. If only the former contribute, the material is a non-conducting dielectric, whereas the latter are predominant in a metal.

For semi-conductors both contributions are important. The former gives rise to the intense absorption to the short wavelength side of the main absorption edge, whilst at long wavelengths free carrier absorption becomes significant.

2.1 DIELECTRICS

In a pure dielectric the wavelength or frequency dependence of the optical constants may be explained simply on the basis of the classical treatment of Lorentz which considers the solid as an assembly of oscillators which are set in forced vibration by the radiation.

We shall discuss the polarization arising from the displacement of bound electrons from their equilibrium positions under the influence of an applied electric field—the field of the incident electromagnetic wave of radiation. Assume a restoring force proportional to the displacement, x, and a damping force proportional to the velocity, i.e. dx/dt. Let the restoring force be $m\omega_0{}^2 x$ and the damping be $mg \cdot dx/dt$. Then the equation of motion of an electron is

$$m \frac{d^2 x}{dt^2} + mg \frac{dx}{dt} + m\omega_0{}^2 x = eE e^{i\omega t} \qquad (2.1)$$

where $Ee^{i\omega t}$ is the impressed electric field.

The solution of this equation shows that x varies sinusoidally at the applied frequency with an amplitude given by

$$x = \frac{eE/m}{\omega_0{}^2 - \omega^2 + i\omega g} \qquad (2.2)$$

Now if there are N electrons per unit volume the polarizability will be $P = Nex/E$ and the dielectric constant will be given by $1 + P/\varepsilon_0$. Hence

$$(n - ik)^2 = \frac{Ne^2/m\varepsilon_0}{\omega_0{}^2 - \omega^2 + i\omega g} + 1 \qquad (2.3)$$

15

giving

$$n^2 - k^2 - 1 = (Ne^2/m\varepsilon_0) \left(\frac{\omega_0{}^2 - \omega^2}{(\omega_0{}^2 - \omega^2)^2 + \omega^2 g^2} \right) \qquad (2.4)$$

and

$$2nk = (Ne^2/m\varepsilon_0) \frac{\omega g}{(\omega_0{}^2 - \omega^2)^2 + \omega^2 g^2} \qquad (2.5)$$

The formulae show that in the neighbourhood of $\omega = \omega_0$ there is an absorption maximum, and that n increases rapidly on decreasing ω through this region. With continuing decrease in ω, n passes through a maximum and falls asymptotically as the frequency gets further away from ω_0. However this asymptotic value is still greater than the value on the short wavelength side of the absorption band, and if there is more than one absorption band n increases as we pass through each towards longer wavelengths.

The more detailed quantum-mechanical treatment of dispersion gives results very similar in form to equations (2.4) and (2.5). (See Rosenfeld, 1951, Mott and Jones, 1936, and Nozieres and Pines, 1958.) This theory shows that the interaction of the field and absorbing atom may be represented by a set of linear oscillators each of which has a resonance frequency corresponding to an allowed transition. The magnitude of the contribution of each oscillator to the optical constants is determined by the oscillator strength f of each transition. Equations (2.4) and (2.5) would then be replaced by a series of terms for all the allowed transitions;

$$n^2 - k^2 - 1 = \sum_j \frac{(Ne^2 f_j/m\varepsilon_0)(\omega_j{}^2 - \omega^2)}{(\omega_j{}^2 - \omega^2)^2 + \omega^2 g_j{}^2} \qquad (2.6)$$

$$2nk = \sum_j \frac{(Ne^2 f_j/m\varepsilon_0)\omega g_j}{(\omega_j{}^2 - \omega^2)^2 + \omega^2 g_j{}^2} \qquad (2.6a)$$

This close similarity between the detailed quantum-mechanical result and the simple classical treatment explains why the latter gives such good agreement with experimental observations on dielectrics.

In principle the sum of the oscillator strengths Σf_j is simply determined from the electronic configuration of the atoms, Seitz (1940), Ditchburn (1952).

It should be noted that the resonant frequency ω_0 is that at which $2nk\omega$, the conductivity, is a maximum. This may be simply verified from equation (2.5) by differentiating $g\omega^2/[(\omega_0{}^2 - \omega^2)^2 + \omega^2 g^2]$. This is contrary to the statement made by Simon (1951) that ω_0

16

represents the maximum of $2nk$. Maxima of k, K and reflectivity also occur at different frequencies from the above.

For a material whose optical properties are represented by a single oscillator and equations (2.4) and (2.5) apply, the parameters may be determined as follows:

(i) ω_0 is ω at maximum $2nk\omega$
(ii) $\frac{1}{2}g$ is $\omega - \omega_0$ * at $2nk\omega = \frac{1}{2}(2nk\omega)_{max}$
(iii) $Ne^2/m\varepsilon_0 = g(2nk\omega)_{max}$

The self-consistency of the theory may then be checked by the fact that at long wavelengths:—

(iv) $n_0{}^2 - 1 = A/\omega_0{}^2$

As $\omega \rightarrow 0$ equation (2.5) shows that the absorption is low while equation (2.4) gives the long wavelength limit of the refractive index

$$n_0{}^2 - 1 = Ne^2f/m\varepsilon_0\omega_0{}^2 \qquad (2.7)$$

An indication of the validity of this equation may be obtained by inserting the data for silicon, namely $N = 5 \cdot 2 \times 10^{22}$ atoms/c.c., $\omega_0 \doteqdot 6 \times 10^{14}$ (from $\lambda_0 = 0 \cdot 5$ μ, Moss, 1952a). Putting $f = 1$ (considering all the absorption spectrum as this one 'line') and giving the other parameters their usual values we obtain $n_0{}^2 \doteqdot 12$, which agrees quite well with the measured value of $n_0 = 3 \cdot 43$ (Briggs, 1950).

For wavelengths much longer than those in the absorption band, k becomes negligible with respect to n, and equation (2.4) may be approximated by

$$n^2 - 1 = (Ne^2/m\varepsilon_0)(\omega_0{}^2 - \omega^2)^{-1} \qquad (2.8)$$

It was formerly considered that a more exact expression for the dispersion was given by use of the Lorentz–Lorenz formula, which is based on the assumption that the effective field polarizing each atom is not simply the average field in the medium but is increased locally. The effect of this theory is to change the left hand side of equation (2.8) to $3(n^2 - 1)/(n^2 + 2)$. This alteration is clearly unimportant if $n \sim 1$, but in semi-conductors where n is high, the difference is significant. Mott and Gurney (1948) have pointed out that experimental results are represented better by the simple formula (2.8) than the Lorentz–Lorenz formula, and Nozieres and Pines (1958) have shown on fundamental grounds that the local field correction should be omitted in highly polarizable solids.

For monatomic semi-conductors, such as germanium, where there

---

* More accurately $g = (\omega_0 - \omega)(1 + \omega_0/\omega)$.

is only one strong absorption band, the dispersion follows the form of equation (2.8) as may be verified by plotting $1/(n^2 - 1)$ against $1/\lambda^2$. The data for germanium, which are plotted in this way in *Figure 10.2*, are seen to lie on a straight line. From the line it is found that $n_0^2 = 15 \cdot 98$ and that ω_0 corresponds to $\lambda_0 = 0 \cdot 48 \ \mu$.

For compounds there is inevitably some degree of ionicity in the crystal, and there must thus be a strong absorption band due to lattice vibration in the far infra-red. In this case the dispersion in the region between the two absorption bands (which is usually the region of most interest) is given approximately by a two term equation similar to (2.8)

$$n^2 - 1 = A_1(\omega_1^2 - \omega^2)^{-1} - A_2(\omega^2 - \omega_2^2)^{-1} \qquad (2.9)$$

where $\omega_1 \gg \omega \gg \omega_2$.

For very ionic crystals the absorption constant becomes so high in the lattice absorption band that the reflectivity approaches 100 per cent. This selective reflection may be used to isolate narrow bands of long wavelength radiation—the so-called 'Restrahlen' bands.

2.2 ARGAND DIAGRAM OF OPTICAL CONSTANTS

It is sometimes informative to plot the optical constants on an Argand diagram. The imaginary part of the dielectric constant $(2nk)$ is plotted vertically; and on the horizontal axis, to the same scale, is plotted the real part minus unity, $(n^2 - k^2 - 1)$. The reasons underlying this type of plot may be seen from a study of the classical dispersion formulae given in section 2.1.

Denote the real part of the dielectric constant by $R + 1$ and the imaginary part by I. Then from equations (2.4) and (2.5), if there is a single resonance frequency ω_0,

$$R = n^2 - k^2 - 1 = \frac{A(\omega_0^2 - \omega^2)}{(\omega_0^2 - \omega^2)^2 + g^2\omega^2}$$

$$I = 2nk = \frac{Ag\omega}{(\omega_0^2 - \omega^2)^2 + g^2\omega^2}$$

where $A = Ne^2/m\varepsilon_0$. Hence $R^2 + I^2 = AR/(\omega_0^2 - \omega^2) = AI/g\omega$. At $\omega = 0$, $R_0 = A/\omega_0^2$. At $\omega = \omega_0$, $I_m = A/g\omega_0$. With some rearrangement it may be shown that

$$I^2 - II_m + R^2 - RR_0\omega_0/(\omega_0 + \omega) = 0 \qquad (2.10)$$

Thus for $\omega \sim \omega_0$

$$(4I - 2I_m)^2 + (4R - R_0)^2 = 4I_m^2 + R_0^2 \qquad (2.11)$$

18

which represents a circle with its centre at $I_m/2$ and $R_0/4$. The upper half of the circle is covered by a range of ω from $\omega_0 - \frac{1}{2}g$ to $\omega_0 + \frac{1}{2}g$: Thus if $g \ll \omega_0$ the $\omega_0/(\omega_0 + \omega)$ term in equation (2.10) does not change much and the curve is almost a circle. The circle cuts the axis at $R = 0$, and $R = \frac{1}{2}R_0$, i.e. at $n^2 - k^2 = \frac{1}{2}(n_0{}^2 + 1)$, and has a diameter equal to the value of $2nk$ at $\omega = \omega_0$.

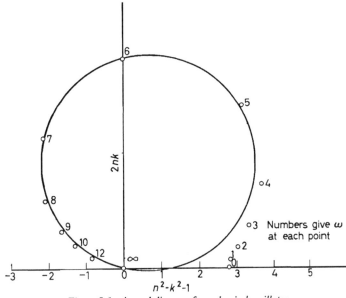

Figure 2.1. Argand diagram for a classical oscillator

Figure 2.1 shows a plot of calculated values of $n^2 - k^2 - 1$ and $2nk$ for various values of ω, when $\omega_0 = 6$, $g = 3$ and $A = 100$. It will be seen that although g is relatively large (i.e. $\sim\omega_0$), most of the points lie near the circle.

Such an Argand diagram for experimental results will show if the optical properties fit the classical dispersion formula with a single resonance frequency. The maximum value of $2nk$ and the resonance frequency ω_0 can readily be obtained from the circle even if no experimental points very close to ω_0 are available.

2.3 THEORETICAL VARIATION OF ABSORPTION AND REFLECTIVITY

As stated in section 2.1, the maximum of $2nk\omega$ occurs at $\omega = \omega_0$. However experimental results do not yield $2nk\omega$ directly; the experimental data to be interpreted is more likely to be the frequency variation of absorption index k (which is readily found from measured values of K) or reflectivity R. It is necessary therefore to study the variation of these parameters in order to be able to derive ω_0 and the other characteristics of the equivalent classical oscillator.

2.3.1 Maximum absorption index, k

The more general form of equation (2.4) is

$$n^2 - k^2 - \varepsilon = (Ne^2/m\varepsilon_0)\left(\frac{\omega_0{}^2 - \omega^2}{(\omega_0{}^2 - \omega^2)^2 + \omega^2 g^2}\right) \qquad (2.12)$$

where we put $\varepsilon = 1$ when dealing with the shortest wavelength absorption band (i.e. the fundamental electronic absorption band), and $\varepsilon = n_1{}^2$, where n_1 is the constant refractive index at frequencies well below the fundamental absorption edge, when we are dealing with an infra-red absorption band such as a Restrahlen band. If $\varepsilon \neq 1$, there are three unknown parameters to be determined to specify the band completely. Making the simplifying substitutions

$$x = \frac{\omega_0{}^2 - \omega^2}{g\omega} = \frac{2(\omega_0 - \omega)}{g} \qquad (2.13)$$

and $\rho = (Ne^2)/(m\varepsilon_0 \omega g\varepsilon)$ we obtain, if $g \ll \omega_0$,

$$n^2 - k^2 = \varepsilon\left(1 + \frac{\rho x}{1 + x^2}\right) \qquad n^2 + k^2 = \varepsilon\left[\frac{(\rho + x)^2 + 1}{x^2 + 1}\right]^{1/2} \qquad (2.14)$$

Hence

$$2k^2/\varepsilon = \left(\frac{(\rho + x)^2 + 1}{x^2 + 1}\right)^{1/2} - 1 - \left(\frac{\rho x}{1 + x^2}\right) \qquad (2.15)$$

Differentiation of this expression (ignoring the slight frequency dependence of ρ) gives the condition for maximum k as

$$4x^3 + 3x^2\rho - 4x - \rho = 0 \qquad (2.16)$$

Of the three possible solutions of this equation the correct one is the small negative value of x, which may be written

$$x = -\sqrt{(x + \rho/4)/(x + 3\rho/4)} \qquad (2.17)$$

For $\rho > 3$ this may be written

$$x = -\sqrt{(\rho - 2)/(3\rho - 2)} \qquad (2.18)$$

so that for large values of ρ, x becomes $-3^{-1/2}$ corresponding to

$$\omega - \omega_0 = 0 \cdot 29g \qquad (2.19)$$

Thus for strong absorption bands the frequency of maximum k is considerably increased from the oscillator resonant frequency. For weak absorption bands where $\rho \ll 1$, equation (2.16) gives $x = -\rho/4$ and the shift is only about half as great. There is no simple expression, independent of ρ, which can be used in conjunction with (2.19) to determine g and ω_0 separately. However, for strong absorption bands an estimate of g can be made using the fact that at $\omega = \omega_0 - g/2$, k has fallen to approximately 40 per cent of its maximum value.

2.3.2 Maximum reflectivity

Using the expressions of equation (2.14) we make the approximation $(\rho + x)^2 \gg 1$. This is valid if ρ is large compared with unity as it transpires that maximum reflectivity occurs at frequencies such that $\rho + x > \frac{1}{2}\rho$.

Rewriting the usual expression for reflectivity (equation (1.24)) as

$$2(R + 1)^2/(R - 1)^2 = (n^2 + k^2 + 1)^2/2n^2 \qquad (2.20)$$

and differentiating with respect to x we find that this expression, and hence R, has a maximum at $x = -0.28\rho + 1/\rho$ for large values of ε. If $\rho^2 \gg 4$ this simplifies to $x = -0.28\rho$ or

$$2(\omega - \omega_0)/g = 0.28\rho \qquad (2.21)$$

Maximum reflectivity thus always occurs on the high frequency side of the resonant frequency, typical values of ρ giving a frequency displacement of one or two bandwidths.

For small values of ε the frequency displacement is increased somewhat. For $\rho^2 \gg 4$ we obtain

$$\varepsilon = 2.5, \qquad 2(\omega - \omega_0)/g = 0.43\rho \qquad (2.22a)$$

$$\varepsilon = 2.0, \qquad 2(\omega - \omega_0)/g = 0.5\rho \qquad (2.22b)$$

The magnitude of the maximum reflectivity may be found by substituting these values into equation (2.20). For the simplest case of large ε and large ρ we obtain

$$(1 + R)/(1 - R)_{\max} = \frac{1}{3}\rho\varepsilon^{1/2} \qquad (2.23a)$$

$$= \frac{1}{3}n_1\rho \qquad (2.23b)$$

This reflectivity function is thus directly proportional to both n_1 and ρ. Use of this expression is probably the most direct way of determining ρ, (assuming n_1 to be known). For InSb for example, where $R_{max} = 78$ per cent (Yoshinaga, 1955*), and $n_1 = 4$, we find $\rho = 6$.

For $\varepsilon = 2$, equation (2.23a) becomes

$$(1 + R)/(1 - R)_{max} = 0 \cdot 6\rho \qquad (2.24)$$

For such materials (e.g. rock salt) very large values of ρ are therefore required to give the very large reflectivities which are observed. For $\rho = 20$, $R_{max} = 85$ per cent, which is about the value observed for rock salt.

Having obtained the value of ρ from the maximum reflectivity in this manner, its substitution in equation (2.21) or (2.22) gives x, the relative frequency displacement from resonance.

The reflectance band is very wide compared with the actual oscillator bandwidth, $\frac{1}{2}g$. For example, for InSb where $\rho = 6$ and $\varepsilon = 16$, (corresponding to $R_{max} = 78$ per cent) the reflectivity has fallen only to 63 per cent at $\omega - \omega_0 = -\frac{1}{2}g$ and $\omega - \omega_0 = 2g$, that is over a frequency range of five half-bandwidths the reflectivity exceeds 80 per cent of its maximum value.

2.3.3 Minimum reflectivity

At a frequency somewhat higher than that for maximum reflectivity, there is a marked minimum of reflectivity. Using the notation of section 2.3.1 and assuming that $x^2 \gg 1$ but that $x + \rho \sim 1$, we find that the minimum occurs when

$$-(x + \rho) = 3^{-1/2} + 2\rho/3\varepsilon \qquad (2.25a)$$

or

$$2(\omega - \omega_0)/g = \rho + 0 \cdot 58 + 2\rho/3\varepsilon \qquad (2.25b)$$

The minimum of reflectivity thus occurs at a frequency several bandwidths above the resonance frequency.

Calculated curves for reflectivity and absorption (k) for a classical oscillator, when $\varepsilon = 14$ and $\rho = 6†$, are shown in *Figure 2.2*.

2.4 INTER-RELATION OF OPTICAL CONSTANTS

It is not generally realized that the two optical constants are not entirely independent. It may be seen however from the classical dispersion equations (2.4) and (2.5) that the real and imaginary parts of the dielectric constant for a given material are both determined at a given frequency by the same parameters N, ω_0 and g. Similarly the detailed quantum mechanical treatment expresses these parts as sums of terms almost identical with equations (2.4)

---

* See section 16.7 for later values.
† These parameters approximate to those of InSb.

and (2.5), each of which is determined by the same parameters f_k, ω_k, g_k for both $n^2 - k^2 - 1$ and $2nk$. It follows that a functional relation must exist between $n^2 - k^2 - 1$ and $2nk$, and if the latter is known *at all frequencies* then the former can be calculated on any frequency.

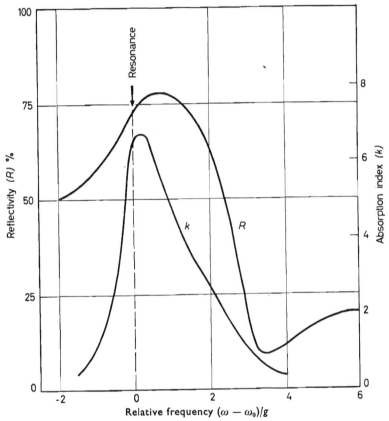

Figure 2.2. *Theoretical frequency dependence of reflectivity and absorption index for a classical oscillator (for $\varepsilon = 14$, $\rho = 6$)*

In addition to relations applying to the dielectric constant, important relations will be given for the reflectivity, and for the refractive index and absorption.

2.4.1 Real and imaginary parts of the dielectric constant

These reciprocal relations may be found by analogy with the

23

electrical formulae given by Bode (1945). For the parameters at an angular frequency a they give:

$$(n^2 - k^2)_a - 1 = \frac{2}{\pi} \int_0^\infty \frac{2nk\omega - a(2nk)_a}{\omega^2 - a^2} \, d\omega \qquad (2.26a)$$

and

$$(2nk)_a = \frac{-2a}{\pi} \int_0^\infty \frac{n^2 - k^2 - (n^2 - k^2)_a}{\omega^2 - a^2} \, d\omega \qquad (2.26b)$$

The first of these relations is derived by summing the dispersion contributions of a series of elementary classical oscillators in Appendix B.1. Also in this Appendix (B.2) it is shown analytically that equation (2.26a) is correct for the dispersion due to a single classical absorption band.

It should be noted that

$$\int_0^\infty \frac{a(2nk)_a}{\omega^2 - a^2} \, d\omega = \int_0^\infty \frac{\text{constant}}{\omega^2 - a^2} \, d\omega = 0$$

so that equation (2.26a) may be written

$$(n^2 - k^2)_a - 1 = \frac{2}{\pi} \int_0^\infty \frac{2nk\omega \, d\omega}{\omega^2 - a^2}$$

and at zero frequency (for a non-conductor where $k_0 = 0$)

$$n_0^2 - 1 = \frac{2}{\pi} \int_0^\infty 2nk \frac{d\omega}{\omega} = \frac{2}{\pi} \int_0^\infty 2nk \frac{d\lambda}{\lambda} \qquad (2.26c)$$

This expression shows that a high value of the low frequency refractive index necessitates a large amount of absorption—i.e. either $2nk$ must be very large or the absorption must extend over a wide relative waveband, or both.

This relation does not provide new information from optical experiments because it is unfortunately not possible in the optical part of the spectrum to determine $2nk$ except by measuring n and k separately or simultaneously. In the radio region, by contrast, $2nk\omega$ is readily measurable as the conductivity. However equation (2.26) is useful as providing an overall check on a complete set of n and k data, and has been used in this way by the author (Moss, 1953b) to check the experimental data for silicon and lead telluride.

The data for PbTe are plotted as $2nk$ in *Figure 14.7*. On replotting as $2nk/\lambda$ and integrating graphically it is found that $n_0 = 5{\cdot}5$, in good agreement with the long wavelength refractive index of $5{\cdot}4$ given by Avery (1951, 1953).

24

Phenomenologically the formula would also indicate that if the total absorption increases—for example by the absorption edge moving to longer wavelengths on cooling—then the refractive index should increase. This behaviour is shown by PbS where the edge does move to longer wavelengths on cooling and dn/dT is negative, whereas for germanium both effects have the opposite sign.

These relationships have been used in the analysis of the optical data for InSb in section 16.8.

2.4.2 Amplitude and phase of reflection

If the complex amplitude of reflection at a surface is represented by

$$r = \sigma e^{-i\phi}$$

where σ, the magnitude, is given by the square root of the reflectivity and ϕ is the phase angle, then

$$\log r = \log \sigma - i\phi$$

The real and imaginary parts of this latter function are related by the same form of equation as $(2.26b)$ namely,

$$\phi_a = -\frac{2a}{\pi} \int_0^\infty \frac{\log \sigma \, d\omega}{\omega^2 - a^2}$$

giving

$$\phi_a = -\frac{a}{\pi} \int_0^\infty \frac{\log R \, d\omega}{\omega^2 - a^2}$$

or

$$\phi_a = \frac{a}{\pi} \int_0^\infty \frac{\log (R_a/R) \, d\omega}{\omega^2 - a^2} \tag{2.27a}$$

This is a particularly important relation, as it implies that a single optical measurement, that of the reflected intensity, made over the whole frequency range, enables the phase angle to be determined for any frequency. From ϕ the optical constants are obtained as follows:

$$r = \sigma e^{-i\phi} = \sigma \cos \phi - i\sigma \sin \phi$$

also

$$r = (n - ik - 1)/(n - ik + 1)$$

Equating real and imaginary parts,

$$k = 2\sigma \sin \phi/(1 + \sigma^2 - 2\sigma \cos \phi)$$

and

$$n = (1 - \sigma^2)/(1 + \sigma^2 - 2\sigma \cos \phi)$$

25

Equation (2.27a) has been used by Jahoda (1957) to calculate the optical constants of barium oxide layers. Also Robinson and Price (1953) have analysed a reflection spectrum. The relation has been applied to the Restrahlen reflection band of ZnS in section 15.5.

Writing equation (2.27a) as

$$2\pi\phi_a = -\int_0^\infty \log R \frac{d}{d\omega}\left(\log\frac{\omega - a}{\omega + a}\right)$$

we can integrate by parts and obtain

$$2\pi\phi_a = -\left[\log R \log\frac{\omega - a}{\omega + a}\right]_0^\infty + \int_0^\infty \log\frac{\omega - a}{\omega + a}\frac{d}{d\omega}(\log R)\,d\omega$$

The term in square brackets is zero, giving

$$2\pi\phi_a = \int_0^\infty \log\frac{\omega - a}{\omega + a}\frac{d}{d\omega}(\log R) \qquad (2.27b)$$

This expression shows that a constant reflectivity term, with zero derivative at all wavelengths, will contribute nothing to ϕ, and thus corresponds to no absorption. In order to use this relation to analyse reflection data it is not necessary to make measurements literally from 0 to ∞ in frequency. To analyse a given absorption band for example, it is only necessary to measure sufficiently far on either side of the band for the reflectivity to be constant over a moderate frequency range, since by virtue of the 'weighting' function $\log (\omega - a)/(\omega + a)$, ϕ_a approaches zero at frequencies well removed from the region of varying R.

2.4.3 *Relation between refractive index and absorption*

The complex polarizability may be written (see equation (2.3))

$$P = (n - ik)^2 - 1 = \beta e^{-i\alpha}$$

where the real and imaginary parts of P are related by equation (2.26). The logarithm of P will also satisfy an equation analogous to (2.27a), namely

$$\alpha_a = \frac{-2a}{\pi}\int_0^\infty \frac{\log\beta\,d\omega}{\omega^2 - a^2}$$

Also

$$r = (n - ik - 1)/(n - ik + 1) = \sigma e^{-i\phi}$$

where

$$\phi_a = \frac{-2a}{\pi}\int_0^\infty \frac{\log\sigma\,d\omega}{\omega^2 - a^2}$$

26

Writing

$$P = (n - ik - 1)(n - ik + 1) = \beta e^{-i\alpha}$$

we obtain

$$(n - ik - 1) = \sigma^{1/2}\beta^{1/2} \exp\left\{-\tfrac{1}{2}i(\phi + \alpha)\right\} \qquad (2.28)$$

and

$$\tfrac{1}{2}(\phi_a + \alpha_a) = \frac{-a}{\pi} \int_0^\infty \frac{\log \beta + \log \sigma}{\omega^2 - a^2}\, d\omega$$

$$= \frac{-2a}{\pi} \int_0^\infty \frac{\log \beta^{1/2}\sigma^{1/2}\, d\omega}{\omega^2 - a^2}$$

Thus the function (2.28) satisfies a relation of the form of equation (2.27a), and so will also satisfy a relation of the form of equation (2.26a). This relation is

$$n_a - 1 = \frac{2}{\pi} \int_0^\infty \frac{k\omega\, d\omega}{\omega^2 - a^2} \qquad \text{or} \qquad n_a - 1 = \frac{2}{\pi} \int_0^\infty \frac{k\omega - ak_a}{\omega^2 - a^2} \cdot d\omega$$

$$(2.29a)$$

with a reciprocal relation of the form of (2.26b) if required.

This relation, which has not previously been published, is potentially an important one as it is quite possible to determine k by a single optical experiment. Transmission measurements are frequently made on thin crystals or layers which give K and hence k. In such measurements a correction for surface reflection—which involves a knowledge of n—is unnecessary if the measured absorption is high enough. Alternatively, in single crystal work, this factor is often eliminated by using specimens of more than one thickness.

Equation (2.29a) is verified analytically for the particular example of free carrier absorption in Appendix B.3.

In terms of the absorption coefficient K, (2.29a) may be written

$$n_a - 1 = \frac{1}{2\pi^2} \int_0^\infty \frac{K\, d\lambda}{1 - \lambda^2/\lambda_a^2} \qquad (2.29b)$$

which becomes, for the zero frequency refractive index,

$$n_0 - 1 = \frac{1}{2\pi^2} \int_0^\infty K\, d\lambda \qquad (2.29c)$$

This expression shows that the long wavelength refractive index is determined simply by the total area under the curve of absorption coefficient against wavelength—it is independent of where in the spectrum the absorption occurs.

27

Transmission measurements of the absorption coefficient of PbTe have been made by Gibson (1950, 1952a) on films and thin crystals from $0\cdot23$ μ to wavelengths beyond the absorption edge. Integrating this data gives $\int_0^\infty K \, d\lambda = 72$ and hence $n_0 = 4\cdot7$. Use of Avery's (1954) data for K gives

$$\int_0^\infty K \, d\lambda = 93 \qquad \text{and} \qquad n_0 = 5\cdot8$$

These values compare favourably with the observed long wavelength refractive index for PbTe of $5\cdot35$ (see *Table 14.1*).

Detailed transmission measurements of the Restrahlen band of rock salt near 60 μ have been published by Barnes *et al.* (1935). These results have been plotted, and give

$$\int_0^\infty K \, d\lambda = 14\cdot3$$

Hence the increase in refractive index on going through this band should be $14\cdot3/2\pi^2 = 0\cdot72$. The experimental difference in optical and radio frequency indices for NaCl is $0\cdot70$ (Mott and Gurney, 1948) in good agreement with this figure.

Equation (2.29a) may be written as

$$n_a - 1 = \frac{-2}{\pi} \int_0^\infty k\omega \, d\left(\log \frac{\omega + a}{\omega - a} \right)$$

Putting $k\omega = \dfrac{\lambda K}{4\pi} \dfrac{2\pi c}{\lambda} = \tfrac{1}{2}Kc$, and integrating by parts,

$$n_a - 1 = \left[\frac{-cK}{\pi} \left(\log \frac{\omega + a}{\omega - a} \right) \right]_0^\infty + \frac{c}{\pi} \int_0^\infty \left(\log \frac{\omega + a}{w - a} \right) \frac{dK}{d\omega} \, d\omega$$

giving

$$n_a - 1 = \frac{c}{\pi} \int_0^\infty \frac{dK}{d\omega} \left(\log \frac{\omega + a}{\omega - a} \right) \, d\omega \qquad (2.29d)$$

since K_∞ is finite.

In this expression the logarithm acts as a 'weighting' function, having a large value when $\omega \sim a$. Hence it is the value of $dK/d\omega$ for $\omega \sim a$ which has most influence on the integral. It follows that n_a will be a maximum when $dK/d\omega$ has a large positive value— i.e. on a long wavelength absorption edge. Such behaviour is seen to occur in the dispersion curves given for ZnS and CdS in Chapter 15.

2.5 METALLIC OR FREE CARRIER ABSORPTION

Absorption by the free, conduction, electrons is of course all-important in a metal, but it is also significant in semi-conductors, particularly at long wavelengths.

In the classical treatment of this problem equation (2.1) is applied, except that as the electrons are free there is no restoring force and hence ω_0 is zero. Equations (2.12) and (2.5) are thus simplified, becoming

$$n^2 - k^2 - \varepsilon = -\frac{Ne^2/m\varepsilon_0}{\omega^2 + g^2} \qquad (2.30a)$$

and

$$2nk\omega = \frac{gNe^2/m\varepsilon_0}{\omega^2 + g^2} \qquad (2.30b)$$

According to quantum-mechanical ideas, electrons moving in the perfectly periodic field of a crystal lattice cannot absorb radiation. In the presence of perturbations due to lattice vibrations, this requirement is relaxed and absorption occurs. Kronig (1931) and Fujioka (1932) have treated the problem as a broadening of the resonance frequency $\omega_0 = 0$ by these perturbations and obtained equations of the same form as (2.30).

From (2.1) we have for a steady field

$$gm \ (\mathrm{d}x/\mathrm{d}t) = Ee$$

But $\mathrm{d}x/\mathrm{d}t = \mu E$ in a steady field, where μ is the mobility at the frequency involved. Hence

$$g = e/\mu m \qquad (2.31)$$

and (2.30b) becomes

$$2nk\omega = cnK = (\sigma_0/\varepsilon_0)/[1 + (\omega\mu m/e)^2] \qquad (2.32)$$

In a semi-conductor the contributions of both bound and free electrons must be considered, and the complete expression for the absorption for example will be given by a series of terms such as (2.6a) plus a term like equation (2.32) for each type of free carrier present. At long wavelengths, well beyond the absorption edge, the contribution from equation (2.6a) will have become negligible and only equation (2.32) need be considered. In addition to the steady absorption given by the equilibrium free carrier densities, it is possible to detect the increase in carrier densities due to injection from point contacts or p–n junctions (Lehovec, 1952; Gibson, 1953) and correspondingly the reduced absorption resulting from carrier extraction (Gibson and Granville, 1957). An elegant experiment has been carried out by Harrick (1956) in which he has studied in detail the spatial distribution of injected or extracted carriers in a bar of germanium by observing the long wavelength absorption.

If $\omega\mu m \gg e$ it follows that K will increase as λ^2. For a typical semi-conductor with $\mu = 0\cdot1m^2/v$.sec. and $m = 10^{-30}$ kg (i.e. approximately the free electron mass) $\omega\mu m/e = 1$ for a wavelength of about 1000 microns, so that the quadratic relation can be expected to hold in general for wavelengths up to a few hundreds of microns. Hence (2.32) becomes

$$K = \frac{e^2\sigma_0}{\varepsilon_0 cn\omega^2\mu^2m^2} = \frac{\lambda^2e^3}{4\pi^2c^3n\varepsilon_0}\left(\frac{N_n}{m_n^2\mu_n}\right) \tag{2.33}$$

for an n type specimen. For near-intrinsic semi-conductors absorption by both types of carrier is important. The original Drude (1900) theory of absorption which postulated two kinds of carrier has been shown by Roberts (1955) to be applicable to our modern concepts of conduction by electrons and holes. The Drude expression may be written as

$$(n - ik)^2 = 1 - \frac{\lambda^2}{2\pi c\varepsilon_0}\left(\frac{\sigma_n}{\lambda_n - i\lambda} + \frac{\sigma_p}{\lambda_p - i\lambda}\right) \tag{2.34}$$

where λ_n and λ_p are characteristic wavelengths given by

$$\lambda_n = 2\pi c\mu_n m_n/e, \qquad \lambda_p = 2\pi c\mu_p m_p/e$$

From this expression we may derive the absorption constant

$$K = \left(\frac{\lambda^2e^3}{4\pi^2c^3n\varepsilon_0}\right)\left(\frac{N_n}{m_n^2\mu_n} + \frac{N_p}{m_p^2\mu_p}\right) \tag{2.35}$$

which may be compared directly with (2.33). This latter equation has been used by Gibson (1956) to interpret absorption in germanium at mm wavelengths.

It is generally found that for wavelengths $\sim$10 μ or longer the mobility μ has approximately its zero frequency value, and thus for such wavelengths equation (2.32) may be used to determine absolute magnitudes. The mass m occurring is of course the effective mass of the electrons (or of the holes for a p-type specimen) and equation (2.32) thus represents a method of calculating this effective mass if the other parameters are measured. This method gives good results for InSb for example (Moss, 1954a,b; Spitzer and Fan, 1957b).

Equation (2.32) may be further generalized to deal with anisotropic effective masses. For one type of carrier in a cubic crystal

30

where the effective masses along the three main axes are m_1, m_2, and m_3, according to Brooks (1955) the mass to be inserted in equation (2.32) is $3/m = 1/m_1 + 1/m_2 + 1/m_3$.

Detailed treatment of the free carrier absorption process shows that equation (2.32) corresponds to the assumption of an energy-independent collision time. If the collision time can be written as depending on the energy by the power law $\tau = \tau_0$ (Energy)$^{-p}$, then it is found (Brooks, 1955) that the absorption as given by equation 2.32 is increased by a factor

$$\gamma(p) = \Gamma(5/2 + p)\Gamma(5/2 - p)[\Gamma(5/2)]^{-2}$$

For $p = 0$ γ is clearly unity; for lattice scattering where $p = \frac{1}{2}$, $\gamma = 1\cdot13$, and for ionized impurity scattering where $p = -3/2$, $\gamma = 3\cdot4$. Little error is thus introduced by ignoring this term (i.e. taking $\gamma = 1$) in intrinsic specimens where lattice scattering will predominate, but in impure specimens the neglect of this factor may be serious.

According to Fan (1956) when the lattice scattering is predominant, the absorption given by equation (2.32) is multiplied by a factor

$$(4/9\pi)(h\omega/kT)^{1/2}[(1 + 2E/h\nu)(1 + E/h\nu)^{1/2}]_{av}$$

where the averaging is done over all the electron energies, E. This energy is of the order of kT for non-degenerate cases, so that if $h\nu \gg kT$ the frequency dependence becomes $\omega^{-1\cdot5}$. At the other extreme of long wavelengths and/or high temperatures, where $h\nu \ll kT$, the law would tend to ω^{-3}. As experiments are usually restricted to the range of photon energies $\frac{1}{2}kT$ to $5kT$ (and for any particular specimen to only part of this range) it is unlikely that any marked divergence from the simple ω^{-2} law would be observed.

Dingle (1955) has shown that the formulae for free carrier absorption are modified slightly by the inclusion of skin effect terms.

Figure 2.3 shows the absorption constant of an n-type sample of InSb plotted against (wavelength)^2. The curves show that the quadratic relation is obeyed for wavelengths from 10 µ upwards.

For very low frequencies where $\omega \mu m/e \ll 1$, then $2nk\omega = \sigma/\varepsilon_0$ and ultimately nk will become very large. For $\sigma = 1\Omega m$ (i.e. about the value for intrinsic germanium) $2nk\omega = 10^{11}$, so that $nk > 1$ if $\omega < 10^{10}$. This shows that only for wavelengths of several cms will pure germanium begin to take on metallic properties. It should be noted that in terms of absorption coefficient we obtain

31

$K = 1635 \ \sigma/n \ db/m$, so that K itself does not increase linearly with wavelength but reaches this constant value.

For impure semi-conductors dispersion effects can be important at wavelengths approaching 100 μ (Fan and Becker, 1951). It is

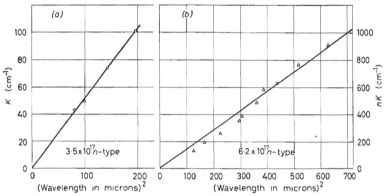

Figure 2.3. Wavelength dependence of free carrier absorption in Insb

then necessary to consider the equation for the real part of the dielectric constant, which will be of the form

$$n^2 - k^2 = n_c{}^2 - \frac{Ne^2/m\varepsilon_0}{\omega^2 + g^2} \qquad (2.36)$$

where n_c is the constant value of n in the non-dispersive region (i.e. ∼3–50 μ for impure Ge). As ω decreases the net effect of equations (2.30) and (2.36) is that the reflectivity falls below its non-dispersive level immediately before rising rapidly to metallic values (Fan and Becker, 1951).

If $\omega^2 \gg g^2$ and $k^2 \ll 1$, then minimum reflectivity will occur when $n \doteq 1$, i.e. when

$$Ne^2/m\omega^2\varepsilon_0 \doteq n_c{}^2 - 1 \qquad (2.36a)$$

Thus for germanium with $\omega = 2 \times 10^{13} \ \mathrm{sec^{-1}}$ (i.e. $\lambda = 100$ μ) we require $N = 5 \times 10^{23} m^{-3}$.

For InSb where the effective mass of the electrons is very small (0·014 free electron masses) the effect is important even in intrinsic material. With $N = 1{\cdot}6 \times 10^{22} m^{-3}$ and $n_c{}^2 = 15{\cdot}4$ we obtain

$$\omega_{\min} = 1{\cdot}5 \times 10^{13} \quad \text{or} \quad \lambda_{\min} = 120 \text{ μ}$$

Yoshinaga and Oetjen (1956) have found a pronounced reflection minimum for pure InSb at just this wavelength.

With any further decrease of ω beyond ω_{min}, equation (2.36) shows that n will quickly approach zero. At the same time it may be seen from equation (2.30b) that k will increase rapidly (approximately as $1/\omega^3 n$), so that the reflectivity quickly reaches high values. Spitzer and Fan (1957) show, for example, that for an InSb sample with $1\cdot2 \times 10^{18}$ cm^{-3} carriers the reflectivity has a minimum ~ 1 per cent at 22 μ, and by 24 μ the reflectivity has reached 70 per cent.

CHAPTER 3

ABSORPTION PROCESSES IN SEMI-CONDUCTORS

3.1 DIRECT AND INDIRECT OPTICAL TRANSITIONS

As INDICATED in Chapter 2, absorption by bound and by free electrons are both significant in semi-conductors. There are alto-gether four types of electrons to be considered: (*1*) inner shell electrons, (*2*) valence band electrons, (*3*) free-carriers—including, of course, holes as well as electrons, and (*4*) electrons bound to localized impurity centres or defects of some type.

The first group do not contribute to either absorption or dis-persion in the spectral regions with which we are concerned, and will not be considered further. Absorption by the second type of electron is of the greatest importance in the study of the funda-mental properties of semi-conductors.

Now in an ideal semi-conductor at zero temperature the valence band would be *completely full*, so that an electron could not be excited to a higher energy state within the band. The only possible absorption is that of quanta sufficiently energetic for the electrons to be excited across the forbidden zone into the empty conduction band. In practice the resulting absorption spectrum is a continuum of intense absorption at short wavelengths, bounded by a more or less steep absorption edge beyond which the material is relatively transparent.

If the limits of the energy bands plotted in $\tilde{k}$ space are as indi-cated in *Figure 3.1a*, i.e. the conduction band minimum and valence band maximum occur at $\tilde{k} = 0$, then the absorption edge may well, though not necessarily* occur at $h\nu = E_g$; where E_g is the minimum width of the forbidden zone or the activation energy of the semi-conductor.

The quantum mechanical selection rule states that if $\tilde{k}_i$ and $\tilde{k}_f$ are the wave vectors of the electron in its initial and final state, and $\tilde{q}$ is the wave vector of the radiation then $\tilde{k}_f - \tilde{k}_i = \tilde{q}$. As for wavelengths of the order of 1 μ or greater $\tilde{q}$ is very small compared with $\tilde{k}$, the selection rule becomes $\tilde{k}_i = \tilde{k}_f$, so that electrons with a given wave number in a particular band can only make transitions to states in a higher band having the same wave number—i.e.

* Whether or not the transition at $\tilde{k} = 0$ is allowed depends on the parity of the initial and final wave functions (Brooks, 1955).

34

in *Figure 3.1a* only 'vertical' transitions are allowed. 'Non-vertical' transitions are nominally forbidden. In practice this does not mean that the latter do not occur at all, but only that absorption due to such transitions is of much lower intensity.

For the case of *Figure 3.1a* the absorption would be intense for all $hv > E_g$ and cease more or less abruptly at $hv = E_g$.

However, if the minimum in the conduction band occurs in a different region of $\bar{k}$ space than the maximum in the valence band

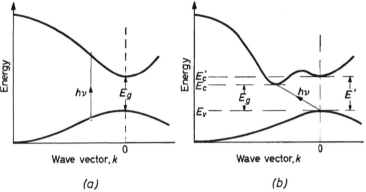

(a)   *(b)*

Figure 3.1. Possible energy bands in semi-conductors

(a band structure as in *Figure 3.1b* for example, which is of the type calculated for germanium by Herman, 1954, 1955*a*) then the *intense* absorption will cease at the wavelength corresponding to the minimum vertical energy gap—i.e. $hv = E'$. Under normal conditions there will always be some factors present which are sufficient to cause relaxation of the selection rules and so permit non-vertical transitions to occur to some degree, momentum being conserved probably by interactions with phonons, so that absorption will in general continue with reduced intensity for frequencies down to $hv = E_g$, where E_g is again the minimum energy gap.

The intensity of the absorption due to direct transitions in the simple case of *Figure 3.1a* is determined primarily by the numbers of occupied states in the valence band and unoccupied states in the conduction which are energetically within hv of each other, and on the transition probability.

According to Fan, (1956) the absorption is given by

$$K = \frac{\pi e^2}{nm} f_{if} N(E) \qquad (3.1)$$

35

where the oscillator strength for the transition is given by

$$f_{if} = \frac{2}{3mh\nu} |M|^2 \tag{3.2}$$

and M is the matrix element governing the transition probability. $N(E)$ is the density of states function. Assume that the energy zones are spherically symmetrical with curvatures corresponding to the effective masses m_e and m_h. From the definition of effective masses we have

$$E - E_g = h^2\tilde{k}^2/2m_e + h^2\tilde{k}^2/2m_h \tag{3.3}$$

As the density of states in $\tilde{k}$ space is simply

$$dN/d\tilde{k} = 2(4\pi\tilde{k}^2) \tag{3.4}$$

we obtain directly

$$\frac{dN(E)}{dE} = 8\pi h^{-3} \left[\frac{2m_e{}^3 m_h{}^3 (E - E_g)}{(m_e + m_h)^3} \right]^{1/2}$$

$$= 8\pi h^{-3} \{2m_r{}^3 (E - E_g)\}^{1/2} \tag{3.5}$$

where the reduced mass $m_r = m_e m_h/(m_e + m_h)$. Assuming that the transition probability is approximately constant in the neighbourhood of the absorption edge (as would seem likely if the transition at $\tilde{k} = 0$ is allowed) then the frequency dependence of the absorption coefficient is determined only by the $(h\nu - E_g)^{1/2}$ term. The magnitude of the absorption coefficient is given by Brooks (1955) as

$$K_d = \frac{e^2 (2m_r)^{3/2}}{\pi n c m_e h^2} (h\nu - E_g)^{1/2} f_{if} \tag{3.6a}$$

and by Bardeen et al. (1956) as

$$K_d = \frac{e^2 (2m_r)^{3/2}}{\pi n c m \hbar^2} (h\nu - E_g)^{1/2} f_{if} \tag{3.6b}$$

Proportionality to $(h\nu - E_g)^{1/2}$ has also been deduced by Dexter (1956). For the case of plane waves, although this may not be too good an approximation, Fan (1956) gives

$$K_d = \frac{2e^2 (2m_r)^{5/2}}{3m^2 \hbar^2 cn} \frac{(h\nu - E_g)^{3/2}}{h\nu} \tag{3.7a}$$

This last expression resembles closely that given by Bardeen, Blatt

36

and Hall (1956) for the case when the transition is not allowed, namely

$$K_d = \frac{2e^2(2m_r)^{5/2}(h\nu - E_g)^{3/2}}{3nc\hbar^2 m_T{}^2 h\nu} \tag{3.8}$$

where m_T is an effective mass for the transition.

If therefore, absorption data can be fitted to a law of type (3.6) it is probable that the absorption is due to allowed direct transitions, whereas higher power laws of $h\nu - E_g$ would be expected for 'forbidden' transitions.

The problem of indirect transitions has been treated by Bardeen, Blatt and Hall (1954, 1956) on the basis that the quantum-mechanical selection rule for conservation of momentum is satisfied by either the absorption or emission of a phonon simultaneously with the photon absorption. The result obtained again depends on whether the transition at $\tilde{k} = 0$ is allowed or not. If it *is* allowed they find

$$K_i \propto (h\nu - E_g)^2 \tag{3.9}$$

and if it is *not* allowed,

$$K_i \propto (h\nu - E_g)^3 \tag{3.10}$$

Fan, Shepherd and Spitzer (1954) point out that with absorption of a phonon the transition rate (and hence K_i) should be proportional to $(\exp \theta/T - 1)^{-1}$, while for phonon emission it should be proportional to $1 + (\exp \theta/T - 1)^{-1}$ or $(1 - \exp - \theta/T)^{-1}$, θ being the Debye temperature. From these proportionality factors and equation (3.9), with the appropriate addition or subtraction of a lattice quantum to the optical quantum, it follows that the indirect absorption should be proportional to

$$\frac{(h\nu - E_g + k\theta)^2}{\exp \theta/T - 1} + \frac{(h\nu - E_g - k\theta)^2}{1 - \exp - \theta/T} \tag{3.11}$$

This expression is variously equated to K by Macfarlane and Roberts (1955b) to $Kh\nu$ by Roberts and Quarrington (1955) and to $K(h\nu)^2$ by Macfarlane and Roberts (1955a). From Bardeen, Blatt and Hall (1954/1956) it would appear that the second of these, i.e. $Kh\nu$, is the preferable one.

In the use of expression (3.11) it is essential that the second term be taken as zero for $h\nu < E_g + k\theta$. With this proviso, a plot of the square root of expression (3.11) against $h\nu$ approximates closely to two straight lines. The line for low values of K cuts the $h\nu$ axis at $h\nu = E_g - k\theta$ and has a slope of $(\exp \theta/T - 1)^{-1/2}$. The line for higher values of K has a slope of $(\coth \theta/2T)^{1/2}$ and (if extrapolated) cuts the axis at $h\nu = E_g + k\theta \tanh \theta/2T$. Either the two intercepts or an intercept and a slope may thus, in principle, be used to find E_g and θ.

This theory seems to represent the low level absorption in Si and Ge reasonably well, but the data for InSb (Roberts and Quarrington, 1955) does not give convincing straight lines.

The general theory of the absorption edge in insulators and semi-conductors, and in particular the effects of lattice imperfections, has been treated in considerable detail by Dexter (1954/6, 1956).

3.2 CHARACTERISTIC FEATURES OF ABSORPTION AND REFLECTION SPECTRA

3.2.1 Absorption spectra

For very few semi-conductors has anything approaching the whole absorption spectrum been measured. Most studies are confined to wavelengths in the neighbourhood of (and beyond) the long wavelength edge, which lies usually in the infra-red. For a few materials the region of maximum absorption—which is usually in the visible part of the spectrum—has been explored, but data at wavelengths much less than the peak (i.e. in the ultra-violet) are very scanty.

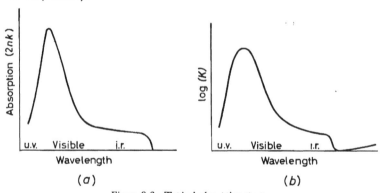

Figure 3.2. Typical absorption spectra

Nevertheless certain typical features are apparent. Figure 3.2a shows characteristic behaviour of $2nk$, the imaginary part of the dielectric constant. The main absorption band will probably be a few eV wide with a peak intensity of $2nk \sim 20$ or 25, and peak absorption constant $\sim 10^6 \text{cm}^{-1}$. Integration over the absorption spectrum in the manner shown by equation (2.26c) indicates that most of the refractive index arises from the absorption in this main band. From the point of view of most photoconductive effects however, it is absorption in the tail band and in the neighbourhood of the absorption edge which is all important. In this band, which does not cover a large range of *energy*, but which may extend over a *wavelength* range of several microns, $2nk$ is commonly of the order of unity and the absorption constant $K \sim 10^3$ or 10^4 cm^{-1}.

The absorption edge of ZnS has been studied by Piper (1953), who finds that for absorption coefficients between 1 mm^{-1} and 10^3 mm^{-1} the measured data lie on a straight line when log K is plotted against energy*. It is significant that the slope of the line is $1/kT$ when natural logarithms are used—i.e. K falls by approximately $e : 1$ for each decrease of kT in the photon energy.

Similar exponential absorption edges have been found in AgBr by Urbach (1953) and Moser and Urbach, (1956). For a wide range of temperatures (116–620°K) these authors find that the plots of log K against frequency (or $h\nu$) are straight lines, with the slope always $1/kT$. Urbach considers that the absorption data for several other materials for example AgCl and CdS, can best be represented by exponential absorption edges, while Aline (1957) finds exponential absorption edges in AgCl†.

In the neighbourhood of the edge the absorption falls to such low values that it becomes difficult to measure the part which is due to band-to-band transitions, as it is liable to be masked by other incidental absorptions or losses in the specimens or experimental equipment. This absorption can however be determined at considerably longer wavelengths by measurement of the photo-conductive effect, as by definition this effect is only produced by that part of the absorption which raises electrons into the conduction band (see Moss and Hawkins, 1958b).

Studies of the spectral distribution of sensitivity of many photo-conductors in the infra-red (see Moss, 1952a) show that there is no absolute limiting wavelength at which photosensitivity vanishes completely, but that the effect generally decreases more or less exponentially with increasing wavelength, and in principle may be extended to any arbitrary limit if sufficiently sensitive measuring equipment is postulated. Correspondingly the band-to-band absorption spectrum cannot have a well defined limit. The photographic sensitivity of AgBr—which similarly depends directly on the band-to-band absorption—has been found to have an exponential fall (Urbach, 1953).

---

* It may be noted that an almost equally straight line is obtained on plotting against wavelength.

† D. L. Dexter (1958) has reported exponential absorption edges in CdS and TlCl.

For materials which show such behaviour it is difficult to specify a unique absorption edge, and yet it is essential to have a unique threshold value of $h\nu$ to compare with the unique energy gap which may be determined by thermal measurements of conductivity or Hall constant. For photoconductors it has been shown by Moss (1952a) that a convenient characteristic wavelength is that at which the photosensitivity falls to half its maximum value (the '$\lambda_{1/2}$' point) and this wavelength will clearly *be near to* the absorption edge. As will be seen, however, from later chapters of this book, it cannot be related specifically to the unique absorption edge.

It would seem likely, *a priori*, that for photon energies only a little greater than the energy gap, the absorption constant will be at least moderately high, and that to seek the smallest measurable K is misleading. This will probably only determine the effect of sundry perturbations or imperfections, for which the probability of transitions taking place may well be given by an exponential function of $1/kT$ in accord with the results quoted above*. This argument could well explain the slow exponential fall in photo-sensitivity which Fan, Kaiser *et al.* (1954) have observed in bombarded Ge. As indicated by equation (3.1) the main para-meter determining the absorption is the density of allowed states $N(E)$. Clearly in an ideal system, with no perturbations on a perfectly regular lattice, $N(E)$ is a discontinuous function, being zero in the case of the conduction band for example when E is below the bottom of the band and finite for E above it. If in practical materials $N(E)$ is not discontinuous*, it would seem reasonable to define the band edge as the point where $N(E)$ changes most rapidly with E, i.e. where $dN(E)/dE$ is a maximum, and correspondingly to define the 'absorption edge' *as the point where the slope of the absorption coefficient is a maximum*†. Such an interpreta-tion has already been suggested by the author (1954a) for InSb.

For a typical absorption spectrum such as that in *Figure 3.2b* there will of course be two such points of maximum steepness, which may correspond to the limits of direct and indirect transitions respectively.

The simplest expression which can be used to represent an absorption curve with an exponential edge is

$$K = K_\infty \{1 + \exp \alpha(E_0 - h\nu)\}^{-1} \qquad (3.12)$$

where α represents the steepness of the edge. The definition of

---

* Parmenter (1955) has shown theoretically that any lattice disorders cause a tailing-off of the density of allowed states curve into the forbidden zone.

† It makes little difference in general whether the point is taken at maximum $dK/d(h\nu)$ or maximum $dK/d\lambda$.

'absorption edge' (ν_0) given above says that $dK/d(h\nu)$ is a maximum, and hence

$$d^2K/d(h\nu)^2 = 0 \qquad (3.13)$$

Equations (3.12) and (3.13) immediately lead to

$$\exp \alpha(E_0 - h\nu_0) = 1 \qquad (3.14)$$

at the 'absorption edge'. Hence for such a curve the parameter E_0 is equivalent to the energy gap and is determined from the point where the absorption has fallen to half its short wavelength value, i.e. at $K = \frac{1}{2}K_\infty$.

Modification of (3.12) by slowly varying functions of $h\nu$ has little effect on the above result. For example, putting

$$K = K_\infty \frac{h\nu}{E_0} \{1 + \exp \alpha(E_0 - h\nu)\}^{-1} \qquad (3.15)$$

we obtain from our definition, using equation (3.13)

$$\exp \alpha(h\nu - E_0) = (\alpha h\nu - 2)/(\alpha h\nu + 2) \qquad (3.16)$$

As $\alpha h\nu \sim \alpha E_0$ the right hand side of equation (3.16) differs little from unity and correspondingly E_0 is very close to $h\nu_0$. For $h\nu_0 = 1$ eV and $\alpha = 1/kT = 40$ eV^{-1}, then equation (3.16) gives more accurately $E_0 = 1 \cdot 002$, so that for normal purposes the error involved in putting $E_0 = h\nu_0$ is negligible. With equation (3.15) the $h\nu_0$ point is determined by extrapolating the *short wavelength* part of the curve, and then finding where the actual absorption coefficient is half the extrapolated value. As by definition the short wavelength part of the curve is only slowly-varying, this can readily be done, and is often simpler graphically than finding the position of maximum slope.

Data for InSb for room temperature are shown plotted in *Figure 3.3*. The experimental points are taken from Moss, Smith and Hawkins (1957). When the residual absorption has been subtracted, the absorption coefficient for $\lambda \geqslant 4$ μ is well represented by the expression:

$$K = K_\infty(h\nu/E_0)^2 [1 + \exp 210(h\nu - E_0)]^{-1} \qquad (3.17)$$

with $E_0 = 0 \cdot 180$ eV. The above value of E_0 is within a few thousandths of an electron volt of the most accurate value for the energy gap of InSb (see section 16.5). It may be noted that data on photoconductivity given by Avery *et al.* (1954) indicate that the exponential fall in the band-to-band part of the absorption continues to photon energies at least as small as $0 \cdot 148$ eV.

41

Such semi-empirical expressions of the type given in equations (3.12), (3.15) and (3.17) may thus provide a good representation of many experimental absorption curves, and if so they may be used to determine a precise value for the energy gap of the material.

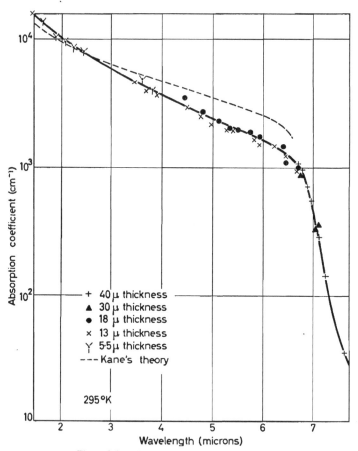

Figure 3.3. Absorption in InSb. (*Moss*, 1957*b;* courtesy *Physical Society, London*)

3.2.2 *Reflection spectra*

Fochs (1956) has studied diffuse reflection spectra from powdered semi-conductors, and finds that as the wavelength is increased a sharp rise in reflectivity is observed in the neighbourhood of the absorption edge.

42

He finds that, in general, the steepest part of the reflectivity/wavelength curve is linear, and uses the onset of this linear region to define the absorption edge. The values of activation energy of several well known semi-conductors determined in this way agree well with those determined by other methods. This method may be particularly useful in the initial evaluation of an unknown material, as the specimen preparation is so simple.

The analysis of specular reflectivity data to give the optical constants is discussed in section 2.4.2.

3.2.3 X-ray spectra

Alternative evidence for the distribution of energy levels in solids is furnished by soft x-ray spectroscopy (Mott and Jones, 1936; Skinner, 1938). Bombardment of the material in an x-ray tube ejects electrons from the inner levels of the atoms of the crystal. These levels are then filled by electrons from the higher bands which emit soft x-ray quanta in the process.

Measurements have been carried out on Se by Givens and Siegmund (1952) and on Si and Ge by Tomboulian and Bedo (1956).

3.3 Pressure and temperature dependence of the absorption edge

One of the theoretical methods of calculating the energy levels in solids is to consider the individual atoms initially in the right relative crystal positions, but widely separated; and then to investigate the behaviour of the wave functions as the atoms are brought closer together and interaction between them increases. As is well known, this interaction causes the discrete levels of the isolated atoms to broaden into bands of allowed energy levels with intervening forbidden zones. *Figure 3.4* shows two possible behaviours of the energy levels. In *Figure 3.4a* the levels broaden continually with reducing lattice spacing, but do not cross, whereas in *Figure 3.4b* the levels cross over. According to Kimball (1935) diamond falls in the latter category.

It is clear from the figures that in the first case (*3.4a*) the energy gap will decrease on compression, whereas for (*3.4b*) the gap will increase on compression, so that the pressure dependence of the activation energy, and hence of the absorption edge, can be of either sign.

For the very small changes in lattice spacing which occur in the accessible pressure range, the change in activation energy with dilatation or pressure may be taken as linear. We may put

$$dE_g \bigg/ \frac{dV}{V} = dE_g \bigg/ \frac{\chi\, dP}{P} = \tfrac{2}{3}(\pm C_e \pm C_h) \qquad (3.18)$$

where V and P are the volume of the specimen and the pressure, χ is the compressibility and C_e and C_h are constants applying to the conduction and valence bands respectively. The alternative signs allow for all possible cases of the bottom of the conduction band and top of the valence band moving either up or down with dilatation.

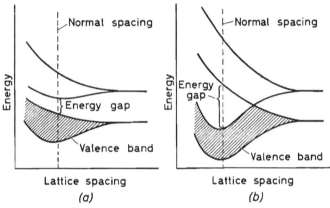

Figure 3.4. Possible energy band developments

Bardeen and Shockley (1950)* developed a deformation theory for the scattering of current carriers, in which lattice waves are considered to set up 'deformation potentials' with which the holes and electrons interact. By this means these workers established the following relation between the constants C_e and C_h and the mobilities in the two bands:

$$C_e^2 \mu_e = 3eh^4 \rho u^2/(512\, k^3 T^3 m_e{}^5 \pi^7)^{1/2} = 7\cdot2 \times 10^{-5} c_{11}(m/m_e)^{5/2} T^{-3/2}$$

$$(3.19)$$

where ρ is the density of the crystal, u the velocity of sound in the crystal, and $c_{11} = \rho u^2$.

The direct shift of absorption edge with pressure has now been observed for one or two materials. Various methods of deducing the energy shift from electrical measurements have been discussed

* The theory is given more fully by Shockley (1950).

by Moss (1952a). More studies have been made however of the shift of the edge with temperature, where there are two main contributing factors. It follows from (3.18) that the dilatation part of the temperature dependence of the energy gap will be given by

$$(dE_g/dT)_a = -(\beta/\chi)dE_g/dP = 2\beta(\pm C_e \pm C_h) \quad (3.20)$$

where β is the coefficient of linear thermal expansion. The second contribution to the temperature dependence arises from an electron-lattice interaction term which is temperature dependent. The problem of the broadening of allowed energy bands by lattice vibrations has been treated by various workers. The qualitative result is that the bands always broaden on heating, so that this part of the energy shift, $(dE_g/dT)_b$, is always negative. Quantitatively Fan (1951) finds

$$(dE_g/dT)_b = -(16k/9\hbar^2 Mu^2)(6\omega^2\pi^2)^{1/3}(m_e C_e{}^2 + m_h C_h{}^2) \quad (3.21)$$

where M is the mass in the unit cell of volume ω. A more recent discussion of this problem has been given by Fan (1956).

Comparison of the measured values of dE_g/dT with the sum of the expressions (3.20) and (3.21) should show which signs are to be used in equation (3.20) and hence give information about the behaviour of the energy bands. Bardeen and Shockley (1950) conclude that for diamond, silicon and germanium both negative signs should be used, while for InSb the indications are that one sign must be positive and one negative—i.e. the two bands move in the *same* direction on compression (see section 16.5).

For the majority of materials the overall shift is to shorter wavelengths on cooling, the most notable exception being the lead chalcogenides. The magnitude of the shift is frequently close to 4×10^{-4} eV/°C.

3.4 DEPENDENCE OF ABSORPTION EDGE ON IMPURITY CONCENTRATIONS

For the vast majority of semi-conductors there is no observable dependence of the intrinsic energy gap on either the type or density of impurities. However, one or two exceptions have now been discovered, notably indium antimonide. For this compound it was found that the absorption edge lay at much shorter wavelengths when it was very *n*-type than if it was intrinsic. Little difference was observed between intrinsic and *p*-type material. The explanation of this effect, given independently by Burstein (1954) and Moss (1954a), is that the energy surfaces in momentum space of the bottom of the conduction band have such marked curvature that the density of states in the conduction band is small, and hence a relatively small density of conduction electrons can fill these states to an appreciable depth and thus affect the absorption edge. It transpires that this curvature is remarkably high in InSb. Also, as the absorption edge for intrinsic material lies at about 7 μ, a shift in the energy gap of only 0·01 eV moves the edge by about one third of a micron, so that very small energy shifts are readily detectable.

In order for this shift with impurity concentration to occur, it is not essential that the lower conduction levels are *completely* occupied, but only that the density of available states corresponding to a given absorption coefficient should occur at a higher energy.

For a material where the energy surfaces are spherical—which appears to be the case for InSb (Pearson and Tanenbaum, 1953)—equation (3.5) gives the density of states directly as a function of energy. For the simple case where only the conduction band has high curvature (i.e. $m_h \gg m_e$) this equation reduces to

$$\frac{\mathrm{d}N(E)}{\mathrm{d}E} = 8\pi h^{-3} \left[2m_e^3(E - E_g)\right]^{1/2} \tag{3.22}$$

Integrating this expression over the energy range 0 to ΔE gives the total number of conduction electrons required to fill the band to a height of ΔE, i.e.

$$N = \int_0^{\Delta E} \mathrm{d}N(E) = (8\pi/3h^3)(2m_e\Delta E)^{3/2} \tag{3.23}$$

Using values of effective mass determined by various methods, see Chasmar and Stratton, 1956; Lax, 1958; Moss, 1958c) equation (3.23) gives values of ΔE for known impurity concentrations which are in good agreement with the measured shift of the absorption edge in n-type InSb. Alternatively the measured shift of the edge may be used to determine m_e (Moss, 1954a). Further theoretical treatment of this problem has been given by Talley and Stern (1955) and Aigrain and des Cloizeaux (1955).

3.5 SHIFT OF ABSORPTION EDGE WITH
MAGNETIC FIELD

It has been established by Burstein, Picus and Gebbie (1956a) that by the use of sufficiently intense magnetic fields, (∼60,000 G) it is possible to produce a measurable movement of the absorption edge in InSb.

The extent of the shift was expected to be approximately

$\frac{1}{2}\hbar\omega_c = Beh/4\pi m^*$ eV. For $m^* = m$ this amounts to only $5 \cdot 6 \times 10^{-9}$ eV/G, but for InSb with its low effective mass of only $0 \cdot 015\ m$, the value is increased to $3 \cdot 8 \times 10^{-7}$ eV/G. This material is also advantageous for such an experiment by virtue of the long wavelengths at which the edge lies. At 7 μ, an energy change of only $2 \cdot 5 \times 10^{-3}$ eV corresponds to a wavelength change of $0 \cdot 1$ μ, which is readily measurable.

Experimentally Burstein, Picus and Gebbie (1956a) found for InSb that the absorption edge shifted to shorter wavelengths with increasing magnetic field to the extent of $2 \cdot 3 \times 10^{-7}$ eV/G.

3.6 INTRINSIC EXCITON ABSORPTION

The treatment of Chapter 3 has so far considered only band-to-band absorption, which by definition generates a free hole and free electron. There is however another possibility, which according to Peierls (1955) is very unlikely in a metal but becomes increasingly more probable as the material studied approaches an insulator, namely that an electron is raised into an excited state but remains bound in a hydrogen-like orbit around either the parent atom or the positive hole from which it came. In the former picture the excitation may be passed on from atom to atom through the crystal, while in the second the bound hole-electron pair may move bodily through the crystal. Both these concepts are usually described as 'excitons' which may move through the crystal. Unless there is separation of the positive and negative charges, no photoconductivity can arise from such absorption. Diemer and Hoogenstraaten (1957) assume that such a process occurs in CdS.

In an ideal crystal therefore, the absorption spectrum should consist of a series of discrete lines corresponding to these exciton states, followed at somewhat higher energies by a continuum which is of course the normal band-to-band, photoconductive, absorption. The theory of exciton absorption has been discussed by Dexter (1954), both from the point of view of the hydrogen-like system immersed in a uniform medium which is the parent crystal, and also on the Heitler–London picture, whence it is found that the overall energy of the system is independent of where the 'excitation' is, so that it may be presumed to pass readily from atom to atom.

The energy levels of the exciton system measured from the beginning of the ionization continuum (i.e. the bottom of the conduction band) will be given by the Bohr formula:

$$E_k = -2\pi^2\mu e^4/h^2\varepsilon^2 k^2 \tag{3.24}$$

where μ is the reduced mass of the electron hole system, i.e.

47

$\mu^{-1} = m_n^{-1} + m_p^{-1}$, and k is the quantum integer. The photon energies of the absorption lines will be given in terms of the energy gap by $h\nu_k = E_g - E_k$.

Such a system of exciton lines has been observed in copper oxide at low temperatures by Gross and Karryev (1952) and Apfel and Hadley (1955) and the observed line frequencies fit the hydrogen-like series formula of equation (3.24) with a high degree of accuracy. Elliot (1957) has given a theoretical discussion of excitons in germanium, and Macfarlane *et al.* (1957, 1958), Zwerdling *et al.* (1958) have observed their effects on absorption spectra. Several papers on excitons, presented at the Rochester Semi-conductor Conference, will be published in *J. phys. Chem. Solids* in 1959.

3.7 CORRELATION BETWEEN ENERGY GAP AND OTHER PHYSICAL PARAMETERS

3.7.1 Compounds

A discussion has previously been given by the author on the relation of the energy gap in compounds to the optical dielectric constant of the material (Moss, 1952a). Based on the general concept that, in a dielectric, energy levels are scaled down by a factor (dielectric constant)^2, the view was put forward that the optical energy gap should vary as n^{-4}, or alternatively that the wavelength of the absorption edge or 'threshold' wavelength of the photoconductive effect would be given by the approximate relation.

$$n^4/\lambda_{1/2} = \text{constant, or } n^4E = \text{constant} \qquad (3.25)$$

It was shown that nearly all the known photoconducting compounds of high refractive index obeyed this law quite well, the value of the 'constant' being 77 for a range of variation of n^4 from 30 to 440.

More recently it has been found by Briggs *et al.* (1954) that the Group III/Group V inter-metallic semi-conductors obey this law satisfactorily with the exception of InSb and InAs. Also Ibuki and Yoshimatsu (1955) have confirmed the agreement for stibnite.

A certain degree of correspondence between activation energies and electronegativities of the constituent ions has been found by Goodman (1955). The III-V compounds show a general dependence of energy gap on melting point (Cunnell and Saker, 1957).

3.7.2 Elements

A general discussion of the correlation between the energy gaps in semi-conducting elements and other physical characteristics has been given by Moss (1952a). A particularly interesting expression which gives approximate values for the activation energies of the diamond, silicon, germanium, gray tin series is derived from equation 3.23 namely

$$E = (h^2/2m)(3N/8\pi)^{2/3} \qquad (3.26)$$

where N is taken as the density of unit cells in the crystal (which equals the density of states in the Brillouin zone). In terms of the normal lattice constant 'a' which is appropriate to four unit cells, $N = 4/a^3$. To obtain the energy gap E_g it is necessary to subtract from (3.26) a 'datum' energy E_0, or alternatively to subtract $N_0 = 4/a_0^3$ from N, i.e.

$$E_g = E - E_0 = \frac{h^2}{2ma^2} \left(\frac{1 \cdot 5}{\pi}\right)^{2/3} - E_0 = \frac{h^2}{2m} \left(\frac{1 \cdot 5}{\pi}\right)^{2/3} \left(\frac{1}{a^2} - \frac{1}{a_0^2}\right)$$

$$(3 \cdot 27)$$

where $E_0 = 2 \cdot 3$ eV or $a_0 = 6 \cdot 59$ Å. With these values and the measured lattice constants, the calculated energy gaps are:—

Table 3.1

	a (Å)	E_g (eV)
Diamond	3·560	5·2
Silicon	5·431	1·02
Germanium	5·657	0·77
Gray Tin	6·46	0·09

The agreement with measured values (see other chapters) is not precise, but the remarkable feature of this expression is that it gives directly the *absolute* values of the widely different activation energies with the use of only one arbitrary constant, a_0. That even this is perhaps not entirely arbitrary is indicated by the fact that it is very near to the value of lattice constant at which the bands cross (and hence gives zero energy gap) in Kimball's (1935) calculation of the band structure of diamond. Alternatively the energy gap *differences* are given with *no* arbitrary constants.

It has been shown by Wolfe et al. (1956) that for the group IV elements and SiC there are correlations between the energy gap, hardness and melting points.

3.8 ABSORPTION BY LOCALIZED IMPURITIES

In an impure semi-conductor, it is possible to excite bound electrons by direct absorption of appropriate optical quanta. For example, in an n-type material with donor levels near the bottom of the conduction band, any electrons which are unexcited due to thermal agitation (i.e. which remain on the donor atoms) may be raised to the conduction band. In many cases—for example group V or group III impurities in Ge or Si—the energy levels are so near to the conduction or valence bands respectively that it is necessary to cool to very low temperatures before the density of impurity centres which are unionized is sufficient to give significant absorption.

Usually the impurity absorption appears as a wide continuum extending to energies as great as that of the main absorption edge. By cooling to liquid helium temperatures however it is possible to narrow the absorption down into discrete lines.

When the energy levels are near to the band edges so that the binding energy is small, the radius of the orbit of an electron on the impurity centre is large compared with the inter-atomic distance and the electron can be considered as moving in the Coulomb field of a charged centre immersed in a medium of dielectric constant equal to that of the crystal. If the effective mass is a simple scalar quantity this problem reduces to the hydrogenic one, with the ionization energy given by

$$E_i = (m_e/m\varepsilon^2)E_h \tag{3.28}$$

where $E_h = 13\cdot6$ eV is the ionization energy for hydrogen.

According to Fan (1956) the absorption coefficient for photo-ionization is given by

$$K = \frac{2^9\pi Nhe^2}{3ncm_e E_i}\left(\frac{E_i}{h\nu}\right)^4 \frac{\exp\left[-4(1-E_i/h\nu)\tan^{-1}(1-E_i/h\nu)\right]}{1-\exp-2\pi(1-E_i/h\nu)} \tag{3.29}$$

At $h\nu = E_i$ this reduces to

$$K/N = 8\cdot3\times10^{-17}m/m^*nE_i \tag{3.30}$$

which for typical semi-conductors gives $K \sim 10^{-16}N$.

3.9 Intra-band absorption

When an energy band is degenerate, as in the case of the valence band of Ge or Si (Herman, 1955a), there is the possibility of optical transitions taking place between the individual bands. If this occurs, then superimposed on the simple λ^2 absorption of the free carriers there will be more or less complex structure resulting from these overlapping energy bands.

Such effects have been observed in the long wavelength absorption spectrum of p-type Ge where it has been found that there are two weak bands with peaks near 3 μ and 4·5 μ (Briggs and Fletcher, 1952) followed by a steep rise to a relatively high absorption level beyond 10 μ (Kaiser, Collins and Fan, 1953). These results, which are shown in *Figure 10.1*, have been interpreted on the basis that the 3 and 4·5 μ absorptions are transitions from the lower band to holes in the two upper bands, and the absorption at 10 μ and beyond as transitions between the two upper bands.

According to Herman (1955a) the structure of the valence band of Si should be similar to that for Ge, but careful measurements up to 12 μ (Fan, Shepherd and Spitzer, 1954) have revealed no detail in the absorption spectrum of *p*-type Si. An absorption band is found in *n*-type however (see section 9.5.3).

3.10 VIBRATIONAL LATTICE ABSORPTION BANDS

In ionic crystals there is a strong absorption band at long wavelengths which arises from the oscillation of the atoms or ions. Although the interatomic forces in ionic crystals are comparable with electronic forces in atoms, the fact that the ionic masses are some 10^4 times greater than electronic masses means that the absorption bands lie at very long wavelengths.

These absorption bands are frequently so intense that the k values are large enough to make the reflectivity high (see section 2.3.2), causing the well known Restrahlen bands.

Such absorption bands occur when there is any degree of ionicity in the binding. Even InSb, where the ionic contribution must be relatively small, has a relatively strong band at about 50 μ (Yoshinaga, 1955; Yoshinaga and Oetjen, 1956).

For very ionic compounds, such as the alkali halides, this Restrahlen absorption band is the main factor determining the dispersion in the region for which these materials are normally used as i.r. prism materials—for NaCl for example in the region 5–15 μ. For these halides the force constant for the oscillating ions appears to be approximately the same for all combinations, and the Restrahlen wavelength (λ_R) is determined mainly by the ionic masses M_1 and M_2. Theoretically it would be expected that $\lambda_R \propto \mu^{1/2}$ where the relation between the reduced mass μ and the ionic masses is $\mu^{-1} = M_1^{-1} + M_2^{-1}$.

In practice the results for the alkali halides seem to be given well by

$$\lambda_R = 14[(M_1 - M_2)/(M_1 + M_2)]^{1/2} \qquad (3.31)$$

The order of reflectivity to be expected may be calculated from the observed absorption coefficients. For rock salt, for example,

51

Barnes *et al.* (1935) found the maximum absorption coefficient to be $K = 1 \cdot 2 \times 10^4$ cm^{-1} at a wavelength of 61 μ. Hence $k = 5 \cdot 7$. This maximum absorption corresponds to $n \sim 1$, so from equation (1.24) $R = 5 \cdot 7^2/(2^2 + 5 \cdot 7^2) = 89$ per cent. Thus although the actual absorption coefficients are not very large, the long wavelengths at which they occur make k, and hence the reflectivity, typical of metallic behaviour.

In the less ionic compounds of groups II/VI and groups III/V, it is not expected that the Restrahlen band will have as much influence on the dispersion as in the alkali halides.

The increase in dielectric constant due to the Restrahlen band is given by equation (16.3).

CHAPTER 4

PHOTOELECTRIC EFFECTS

4.1 GENERAL DESCRIPTION

WITH semi-conductors it is possible to observe both external and internal photo-effects. In the former case electrons are liberated entirely from the material into a surrounding vacuum and collected by an anode. As already described by the author (Moss, 1952a) the phenomenon is essentially similar to that occurring at metallic photocathodes, although significant differences are found between the retarding potential curves of semi-conductors and metals. In spite of the attention paid to this effect by Apker and his co-workers (1948, 1949 and 1954) and by Arseneva-Gail (1949, 1955) it remains of much less importance than the internal photoelectric effect, since the amount of fundamental information it yields is very limited and essentially rather qualitative in nature.

In the internal photo-effect the primary process is the absorption of a photon of sufficient energy to excite an electron into the conduction band.

If the electron originates in the full band then intrinsic photo-conductivity results, and both holes and electrons contribute to the photocurrents. The spectral distribution of the effect is determined essentially by the absorption spectrum of the pure semi-conductor, the long wavelength limit of sensitivity lying in the neighbourhood of the absorption edge. In extrinsic photoconductivity either electrons are excited from donor levels into the conduction band giving electron conduction, or from the full band to acceptor levels giving hole conduction. In either case the photoconductive limit is governed by the relevant impurity activation energy.

For descriptive introductory treatments of internal photo-effects see Moss (1952a), Rose (1955).

4.2 INTERNAL PHOTO-EFFECTS

The photo-effect may be manifested or observed in a variety of different ways, for example: (1) with no applied fields—photo-voltaic effects at point contacts or p–n junctions, and Dember effect; (2) with applied electric field—photoconductivity; and (3) with applied magnetic field—production of the photoelectro-magnetic effect, or variation of Dember e.m.f. Also direct production of photo-motive forces.

All the effects arise from the primary process of light injection of additional carriers, the magnitude of the measured parameters depending on the external factors such as the intensity and wavelength of the irradiation and applied electric and magnetic fields, and on internal factors such as specimen geometry, absorption constant, mobility and recombination processes.

The most important of the fundamental equations governing the photo-effects are a continuity equation which simply states that at equilibrium the total rate of generation of holes and electrons equals the total rate of recombination, and a set of electric and magnetic field equations to describe the motions of the carriers under the combined action of these fields and diffusion. In addition there are subsidiary equations or assumptions regarding neutrality or space charges, surface recombination laws, Fermi levels, dependence of diffusion constants on mobility and carrier concentration etc. The general equations are:—

$$\partial p/\partial t = q - (p - p_0)/\tau - (1/e) \operatorname{div} \tilde{J}^+ \qquad (4.1)$$

$$\partial n/\partial t = q - (n - n_0)/\tau + (1/e) \operatorname{div} \tilde{J}^- \qquad (4.2)$$

$$\tilde{J}^+ = e\mu_h pE - eD_h \operatorname{grad} p \qquad (4.3)$$

$$\tilde{J}^- = be\mu_h nE + beD_h \operatorname{grad} n \qquad (4.4)$$

$$\tilde{J} = \tilde{J}^+ + \tilde{J}^- \qquad (4.5)$$

$$\operatorname{div} \tilde{E} = 4e(\Delta p - \Delta n)\varepsilon \qquad (4.6)$$

where J^+ and J^- are the hole and electron currents respectively, $\Delta p = p - p_0$ and $\Delta n = n - n_0$ are the excess carrier densities over their equilibrium values, μ_h and D_h are the mobility and diffusion constants of holes, b is the mobility ratio, E the electric field, q the rate of quanta absorption (which is taken as the rate of generation of hole-electron pairs) and τ the carrier lifetime. For normal semi-conductors it is a good approximation to assume that charge neutrality exists throughout the specimen, which means $\Delta n = \Delta p$ everywhere, (the effect of not making this assumption is seen later in a particular case in section 4.2.1). The first four equations may now be solved for Δp, J^+, J^- and E, equation (4.6) being redundant. Eliminating the Js we obtain:

$$\partial \Delta p/\partial t = q - \Delta p/\tau + D_h \operatorname{div} \operatorname{grad} \Delta p - \mu_h p \operatorname{div} \tilde{E} - \mu_h \tilde{E} . \operatorname{grad} \Delta p$$
$$(4.7)$$

$$\partial \Delta n/\partial t = \partial \Delta p/\partial t = q - \Delta p/\tau + bD_h \operatorname{div} \operatorname{grad} \Delta p$$
$$+ b\mu_h n \operatorname{div} \tilde{E} + b\mu_h \tilde{E} . \operatorname{grad} \Delta p \qquad (4.8)$$

The corresponding equations when trapping terms are included have been given by Rittner (1956).

In the small-signal case, where $\Delta p \ll p$ or n and we can put p_0 and n_0 for p and n in the fifth terms, the equations become linear and yield

$$\partial \Delta p / \partial t = q - \Delta p / \tau + D \operatorname{div} \operatorname{grad} \Delta p + \mu \tilde{E} \cdot \operatorname{grad} \Delta p \quad (4.9)$$

with

$$\left. \begin{array}{l} D = \mu_e \left[(n_0 + p_0)/(bn_0 + p_0) \right] \\ \mu = D_e \left[(p_0 - n_0)/(bn_0 + p_0) \right] \end{array} \right\} \quad (4.10)$$

where D is the ambipolar diffusivity and μ is an effective mobility.

Equation (4·9) is still too complex to be solved in general, and further restrictions must be imposed in individual problems. By such means very useful solutions of the equations can be obtained, in particular steady state solutions result from putting $\partial \Delta p / \partial t = 0$ and for the intrinsic case $\mu = 0$. Some of the interesting applications of this theory will now be treated in more detail.

4.2.1 Intrinsic material

We will treat the case of an intrinsic material where $n_0 = p_0 = n_i$ so that μ is zero, with the earlier proviso of small signal theory. Charge neutrality will not be assumed initially. We will treat the configuration of a rectangular block whose length and width are very large compared with the thickness t. Length and width are taken in the x and z directions, and the specimen is uniformly illuminated in the y direction. End effects are neglected and carrier distributions are assumed to be uniform over the length and width so the problem becomes one-dimensional. (An example where the ends of the specimen are in the dark has been treated by Rittner, 1956.) The net current flow in the y direction at any point will be zero, so that

$$J_y^+ + J_y^- = 0 \quad (4.11)$$

If the illumination is such that the rate at which quanta pass into the specimen is I per unit area, then the rate of carrier generation at any depth is

$$q = KI \exp(-Ky) \quad (4.12)$$

From equations (4.1)–(4.6) we obtain:—

$$J_y^+ = -eD_h \, d\Delta p/dy + e\mu_h E_y n_i \quad (4.13)$$

$$J_y^- = -J_y^+ = ebD_h \, d\Delta n/dy + eb\mu_h E_y n_i \quad (4.14)$$

$$dJ_y^+/dy = eKI \exp(-Ky) - e\Delta p/\tau \quad (4.15)$$

$$\Delta n - \Delta p = -(\varepsilon/e) \, dE_y/dy \quad (4.16)$$

5

Eliminating Δn, E_y and $J_y{}^+$ gives for Δp,

$$\frac{d^4\Delta p}{dy^4} - \frac{d^2\Delta p}{dy^2} + K^3 I\tau \exp(-Ky) =$$
$$\frac{(b+1)\,e\mu_h n_i}{2\varepsilon b D_h} \left[L^2\!\left(\frac{d^2\Delta p}{dy^2}\right) - \Delta p + KI\exp(-Ky) \right] \quad (4.17)$$

where $L = (D\tau)^{1/2}$ is the ambipolar diffusion length.

Now the factor $e\mu_h n_i/\varepsilon D_h$ is extremely large ($\sim 10^{13}$ for Ge) so that we can ignore all terms which do not contain this factor and simply solve the right hand side of equation (4.17) i.e.

$$L^2\,d^2\Delta p/dy^2 - \Delta p = \overset{\smile}{K} I\tau \exp(-Ky) \quad (4.18)$$

This same equation results from equations (4.13), (4.14) and (4.15) if we put $\Delta n = \Delta p$—i.e. make the assumption of charge neutrality as suggested in section 4.1. This confirms that such an assumption is generally a good approximation. The solution of equation (4.18) is

$$\Delta p = P_1 \cosh y/L + P_2 \sinh y/L + \frac{KI\tau \exp(-Ky)}{1 - K^2 L^2} \quad (4.19)$$

with the boundary conditions

$$y = 0, \quad J_y{}^+ = -es\Delta p; \qquad y = t, \quad J_y{}^+ = es\Delta p \quad (4.20)$$

if s, the surface recombination velocity, is the same on both front and back surfaces.

4.2.2 Photo-voltaic effects

When an experimental measurement is made of the photo-voltage generated at a probe placed on an illuminated surface, what is determined fundamentally is the voltage difference between the Fermi level at some distant, unilluminated, part of the specimen where conditions are those of thermal equilibrium, and the pseudo-Fermi level in the region immediately under the probe.

For the densities of holes and electrons in the equilibrium region we have

$$n_0/N_c = \exp -(e_c - \eta_0) \qquad p_0/N_v = \exp -(\eta_0 - e_v) \quad (4.21)$$

where η_0 is the Fermi level and e_c and e_v are the energy levels of the bottom of the conduction band and top of the valence band (all in units of kT), and N_c and N_v are the densities of states in the conduction and valence bands. Hence

$$\log n_0/p_0 = 2\eta_0 - e_v - e_c + \log N_c/N_v \quad (4.22)$$

This expression shows of course that if $n_0 = p_0$ and $N_c = N_v$ then the Fermi level lies midway between the bands. Similarly in the illuminated region, if η is the position of the pseudo-Fermi level

$$\log n/p = 2\eta - e_v - e_c + \log N_c/N_v \qquad (4.23)$$

Putting $p = p_0 + (\Delta p)_0$, $n = n_0 + (\Delta p)_0$, $\eta = \eta_0 + \Delta\eta$, where $(\Delta p)_0$ is the value of Δp at the surface $y = 0$, then we find that

$$\text{Photo-voltage} = \Delta\eta = (\Delta p)_0(p_0 - n_0)/2p_0 n_0$$
$$= (\Delta p)_0(p_0 - n_0)/2n_i{}^2 \qquad (4.24)$$

This equation shows that the photo-voltage reverses sign as the material goes from p-type to n-type, and would be zero for a truly intrinsic material. However the voltage observed is directly proportional to $(\Delta p)_0$ however small $p_0 - n_0$ is, so that the equations of section 4.2.1 can be used to study the behaviour of the photo-voltage in conditions where the material is approximately intrinsic, but departs sufficiently from the truly intrinsic condition to provide observable signals. (It may be noted in the best possible germanium or indium antimonide $(p_0 - n_0)/n_i$ is at least several per cent.)

Putting the boundary conditions of equation (4.20) into equation (4.19), and making the assumption of a thick specimen so that we may put $\cosh t/L \doteqdot \sinh t/L \gg 1$ and $\exp(-Kt) = 0$ we obtain

$$\Delta p = \frac{KI\tau}{K^2L^2 - 1} \left[\left(\frac{KL + \alpha}{1 + \alpha} \right) (\cosh y/L - \sinh y/L) - 1 \right] \quad (4.25)$$

where the dimensionless parameter

$$\alpha = \tau s/L \qquad (4.26)$$

On the surface we have

$$(\Delta p)_0 = KI\tau/(KL + 1)(\alpha + 1) = \frac{I}{s}\left(\frac{\alpha}{1 + \alpha}\right)\left(\frac{KL}{1 + KL}\right) \quad (4.27)$$

If the surface is well etched so that s is small and $\alpha \ll 1$, then

$$(\Delta p)_0 = I\tau K/(KL + 1) \qquad (4.28)$$

Alternatively if s is very large so that $\alpha \gg 1$, equation (4.27) becomes

$$(\Delta p)_0 = (I/s)KL/(KL + 1) \qquad (4.29)$$

At short wavelengths where K is very large the photo-voltage will be a maximum determined for any α by

$$(\Delta p)_{0\,\text{max}} = I\alpha/s(1 + \alpha) \qquad (4.30)$$

so that the photo-voltage falls to half its maximum value when

$$KL = 1 \qquad \cdot \qquad (4.31)$$

This relation gives an important way of determining the diffusion length L (and hence τ, assuming the mobilities to be known). It is necessary to measure the spectral sensitivity curve, determine the $\lambda_{\frac{1}{2}}$ point, and then from prior knowledge or a subsidiary experiment evaluate the absorption constant K at this point. In contrast with the simpler methods of measuring diffusion length this has the merit that it does not measure it *along* the surface where surface imperfections may give spurious values, but down into the body of the semi-conductor. Furthermore the lifetime derived is the actual bulk lifetime, irrespective of the extent of surface recombination, which (as shown by equation 4.29) merely reduces the magnitude of the response equally at all wavelengths. This method has been used by Moss (1955a) to determine the lifetime in InSb—giving values which are reasonably consistent with those obtained by other methods—and in PbSe crystals, where it is the only method so far known to have yielded values at all.

As stated at the beginning of this section, a measurement of the photo-voltaic effect gives the shift of the Fermi level on illumination. An interesting alternative way of observing this shift was devised by Wlerick (1954) who illuminated strongly *half* a layer of cadmium sulphide and then showed that the system acted as a rectifier with an internal potential barrier. Bulk photo-voltaic phenomena have also been observed by Tauc (1955).

4.2.3 *Photoconductivity in intrinsic material*

Photoconductivity of the illuminated specimen may be measured by the use of electrodes to which is applied a small field E_x in the x direction. Such a field will not disturb significantly the uniform carrier distributions already postulated for the x and z directions or the distribution in the y direction given by equation (4.19).

The photocurrent per unit cross sectional area will be given directly in terms of Δp and the mobilities by:—

$$\Delta i = \frac{e(b+1)\mu_h}{t} E_x \int_0^t \Delta p \, dy \qquad (4.32)$$

or

$$\Delta i \propto \Delta P \quad \text{where} \quad \Delta P = \int_0^t \Delta p \, dy \qquad (4.33)$$

Solving equations (4.19) and (4.20) without the approximations used in deriving equation (4.25) we obtain

$$\Delta P(1 - K^2 L^2)/I\tau = 1 - \exp(-Kt) +$$
$$KL(\alpha \cosh t/L + \sinh t/L - \alpha) \left(\frac{(KL - \alpha) \exp(-Kt) - KL - \alpha}{(1 + \alpha^2) \sinh t/L + 2\alpha \cosh t/L} \right)$$
$$(4.34)$$

It may be noted that by detailed expansion the right hand part of this equation can be shown to be zero at $KL = 1$ so that ΔP is finite at this point.

This equation is rather complex and it is interesting to study its form in various specific cases:

(1) Zero surface recombination velocity, i.e. $\alpha = 0$, we find

$$\Delta P/I\tau = 1 - \exp(-Kt) \qquad (4.35)$$

or

$$\Delta P = \tau \text{ (total absorbed quanta)}$$

which is the form given by simple photoconductivity theory.

(2) Thick samples, small recombination velocity.

Assuming $t \gg L$ and $\alpha \sim 1$ so that $KL\alpha \gg 1$ at short wavelengths we obtain for ΔP_∞, the number of excess carriers as $K \to \infty$.

$$\Delta P_\infty = I\tau/(1 + \alpha) \qquad (4.36)$$

This equation may be written as $\Delta P_\infty = I\tau_{\text{eff}}$ where the effective time constant is given by

$$1/\tau_{\text{eff}} = 1/\tau + s/L \qquad (4.37)$$

As lifetimes are customarily added reciprocally we see that the surface recombination velocity has the effect of a 'surface recombination time, L/s'.

(3) Thick samples, high surface recombination.

For the case where $\alpha \gg 1$, but where $\alpha \ll KL$ at short wavelengths, we have

$$\Delta P_\infty = I\tau/\alpha \qquad (4.38)$$

which of course goes to zero as s and α become infinite.

(4) Infinite surface recombination velocity.

Even when $\alpha = \infty$, ΔP is not zero for all values of K, but rises to a maximum in the neighbourhood of the absorption edge. For $\alpha = \infty$ equation (4.34) becomes

$$\Delta P(1 - K^2 L^2)/I\tau =$$
$$1 - KL \tanh t/2L - (1 + KL \tanh t/2L) \exp(-Kt) \quad (4.39)$$

This expression is found to have a maximum at $KL = 0.54$ for a specimen five diffusion lengths thick (i.e. $t/L = 5$). The maximum response is then given by

$$\Delta P_{max} = 0.51\tau I \qquad (4.40)$$

a surprisingly high value considering that the recombination velocity has been taken as infinite on both surfaces of a specimen only a few diffusion lengths thick.

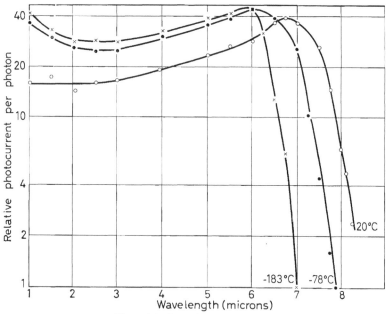

Figure 4.1. Spectral sensitivity of InSb

A method of measuring α, and hence s, is to determine the relative values of ΔP_{∞} and ΔP_{max} from a spectral sensitivity curve (which should be plotted as quantum sensitivity, not energy sensitivity). From equations (4.38) and (4.40) we have (for a specimen five diffusion lengths thick):

$$\Delta P_{max}/\Delta P_{\infty} = 0.51\alpha \qquad (4.41)$$

Figure 4.1 is a spectral response curve for InSb which shows a significant peak in the neighbourhood of the absorption edge. This specimen was approximately five diffusion lengths thick. From the curve $\Delta P_{max}/\Delta P_{\infty} = 2.4$ and hence $\alpha = 4.7$. For this

specimen $D = 36$ cm²/sec and $\tau \sim 10^{-7}$ sec. Since $s = L\alpha/\tau$
$= \alpha(D/\tau)^{1/2}$ we find $s \sim 15 \times 10^4$ cm/sec.

For thinner specimens the spectral response curve becomes
even more peaked in the neighbourhood of the absorption edge.
When $t = L$ the peak occurs at $KL = 2\cdot6$. Taking $L = 40$ μ
(typical of intrinsic InSb) the values of $\Delta P/I$ at various absorption
levels are:—

$$K(\text{cm}^{-1}) \quad 10^5 \quad 10^4 \quad 10^3 \quad 10^2 \quad 10$$
$$\Delta P/I\tau(\%) \quad 0\cdot12 \quad 1\cdot2 \quad 6\cdot3 \quad 21 \quad 0\cdot2$$

Such a variation in K can well take place within a narrow wavelength range near the absorption edge. Marked peaks in the
spectral sensitivity curve comparable with the above figures
are often observed in CdS for example, and have been reported
for InSb by Frederikse and Blunt (1956).

The theory of the intensity dependence of the photocurrent at
high illumination levels is given in Appendix C.

4.3 THE DEMBER OR PHOTO-DIFFUSION EFFECT

If the surface of a semi-conductor is illuminated with strongly
absorbed radiation, a high concentration of hole-electron pairs
will be produced near the illuminated surface. A concentration
gradient will be set up and the carriers will diffuse away from the
surface. Assuming the electrons to be more mobile (as is almost
always true) they will diffuse more rapidly than the holes and will
set up a negative charge on the back (unilluminated) surface. The
field set up by this charge will oppose the electron flow and assist
the hole flow so that in equilibrium there will be net current flow
through the sample. The photo-diffusion voltage set up in this way
is usually called the Dember e.m.f. after its original discoverer.
(See Dember, 1931.)

To illustrate the main features of the Dember effect we shall
treat the case of intense absorption, which will be represented
analytically by a surface generation term in the boundary conditions,
in lieu of the previous bulk generation term. This case will thus
be representative of the effect at wavelengths well below that of
the absorption edge.

The configuration considered is the same as in section 4.2.1—i.e.
a plane parallel slab illuminated normal to the surface in the y
direction. We again make use of the neutrality condition, $\Delta n = \Delta p$.

The problem reduces to a one-dimensional one, no fields are
set up in the x or z directions, and the maximum concentration
gradient lies along the y axis. The photo-diffusion voltage is

measured between the illuminated surface $y = 0$ and the dark surface at $y = t$. We no longer assume that the recombination velocity is the same on both surfaces, or that the material is intrinsic. The relevant equations for the small signal condition are

$$J_y^+ = p_0 e \mu_p E_y - e D_p \, d\Delta p / dy \qquad (4.42)$$

$$J_y^- = b n_0 e \mu_p E_y + e D_n \, d\Delta p / dy \qquad (4.43)$$

$$J_y^+ = -e \Delta p / \tau \qquad (4.44)$$

The Dember e.m.f. is of course an open circuit voltage so no net current flows in the y direction. Hence $J_y^+ + J_y^- = 0$.

Eliminating the Δp term from equations (4.42) and (4.43) gives,

$$E_y = \{(b - 1)/[b e \mu_p (n_0 + p_0)]\} \cdot J_y^+ \qquad (4.45)$$

Substituting in equation (4.42)

$$J_y^+ = -e[(n_0 + p_0)/(b n_0 + p_0)] D_n \, d\Delta p / dy \qquad (4.46)$$

We see from this equation that the effect of the field E_y has been to modify the diffusion conditions so that the holes (and also the electrons) appear to diffuse with a diffusion constant given by

$$D = (n_0 + p_0) D_n / (b n_0 + p_0) \qquad (4.47)$$

This is the ambipolar diffusion constant which has already been used in section 4.1.

Eliminating Δp and E_y we have

$$d^2 J_y^+ / dy^2 = J_y^+ / L^2 \qquad (4.48)$$

which with the boundary conditions

$$y = 0, \quad J_y^+ = eq - e s_1 \Delta p; \qquad y = t, \quad J_y^+ = e s_2 \Delta p \qquad (4.49)$$

gives

$$J_y^+ = eq \, \frac{\sinh (t - y)/L + \alpha_2 \cosh (t - y)/L}{(1 + \alpha_1 \alpha_2) \sinh t/L + (\alpha_1 + \alpha_2) \cosh t/L} \qquad (4.50)$$

where

$$\alpha_1 = s_1 \tau / L \qquad \alpha_2 = s_2 \tau / L \qquad (4.51)$$

The photo-diffusion voltage is given by

$$V_D = E_t - E_0 = \int_0^t E_y \, dy = \frac{b - 1}{e \mu_n (n_0 + p_0)} \int_0^t J_y^+ \, dy \qquad (4.52)$$

and

$$\int_0^t J_y^+ \, dy = eqL \, \frac{[\cosh t/L + \alpha_2 \sinh t/L - 1]}{(1 + \alpha_1 \alpha_2) \sinh t/L + (\alpha_1 + \alpha_2) \cosh t/L} \qquad (4.53)$$

For the case of a thick specimen where $\cosh t/L \sim \sinh t/L \gg 1$ this reduces to

$$V_D = \frac{(b-1)qL}{\mu_n(n_0 + p_0)(1 + \alpha_1)} \qquad (4.54)$$

This equation makes clear the general behaviour of the Dember voltage. We see that: (1) it is proportional to the irradiation intensity q (for the small signals considered); (2) it is proportional to the diffusion length; (3) it depends directly on the *difference* in mobilities, $b - 1$. It vanishes if the electrons and holes have the same mobility; (4) its magnitude is reduced if the recombination velocity on the illuminated surface, and hence α_1, is high. For the thick specimen approximation used in equation (4.54) it is of course independent of α_2 and s_2; and (5) the effect is inversely proportional to $n_0 + p_0$, so that for a given material it is a maximum at $n_0 = p_0$, i.e. for intrinsic material.

It should be mentioned that this treatment of the Dember effect ignores the presence of surface states. Such surface states, and the variation of surface potential on illumination (Brattain, 1951), may make it difficult to measure the Dember effect experimentally.

The change in the Dember e.m.f. caused by a transverse magnetic field has been observed by Aron and Groetzinger (1955) and has been studied both theoretically and experimentally by Walton and Moss (1958).

4.4 THE PHOTOELECTRO-MAGNETIC (P.E.M.) EFFECT

4.4.1 Introduction

The photoelectro-magnetic (P.E.M.) effect (which is also variously described as the photo-magneto-electric or magneto-photo-voltaic effect) was originally discovered in cuprous oxide by Kikoin and Noskov (1934). The effect was 'rediscovered' independently by Moss, Pincherle and Woodward (1953) and by Aigrain and Bulliard (1953), both groups working with germanium. Its use in the investigation of lifetime and surface recombination in germanium has been reported by Bulliard (1954a), Aigrain (1954a), Oberly (1954), Grosvalet (1954) and Buck and Brattain (1955). The effect has also been studied in PbS by Moss (1953a), in Si by Bulliard (1954b), in Cu_2O by Komar et al. (1954) and in InSb by Kurnick et al. (1954) and by Kurnick and Zitter (1956a, b). Papers on the effect have recently been published by Pincherle (1956), Lagrenaudie (1956), Garreta and Grosvalet (1956), Moss (1956/8), and Gartner (1957).

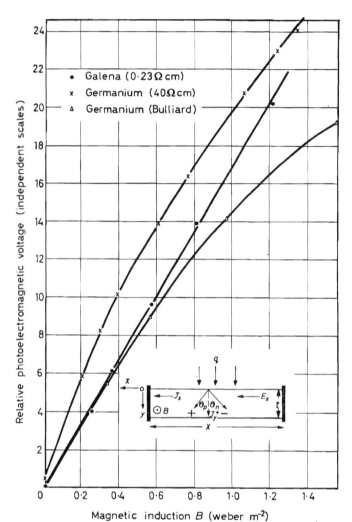

Figure 4.2. Field dependence of P.E.M. effect

The P.E.M. effect may be described as the Hall effect of a photo-diffusion current. In *Figure 4.2* radiation is falling perpendicularly onto a slab of photoconductor. Electron-hole pairs are produced at the illuminated surface, set up a concentration gradient, and diffuse downwards. Under the influence of the magnetic induction B, which is perpendicular to the plane of the

64

diagram, the electrons and holes separate and there is a net charge transport along the sample. If the ends of the sample are short-circuited, a steady current will flow; if the ends are open-circuited, a voltage difference will build up until it is just sufficient to prevent current flow.

4.4.2 Simple phenomenological theory

An approximate treatment of the short-circuit current may be given as follows. Consider small signal and small magnetic field conditions so that equilibrium values of the parameters may be employed. In the configuration of *Figure 4.2* suppose q quanta per second per unit area are absorbed in an intrinsic specimen, and that each quantum produces a hole-electron pair. As a result of the requirement that no current flows down through the slab, the diffusion lengths for both carriers must be equal, i.e. $L_n = L_p = \sqrt{D\tau}$, where D is the ambipolar diffusion constant. Charge transfer between the electrodes per electron-hole pair is

$$(D\tau)^{1/2} e(\tan \theta_n + \tan \theta_p)/X$$

where θ_p and θ_n are the Hall angles for holes and electrons respectively, and the short-circuit current (per unit width of specimen in the magnetic field direction) is;

$$J_{sc} = qe(D\tau)^{1/2}B(\mu_n + \mu_p) = qB[2kT\tau e \,\mu_n\mu_p(\mu_n + \mu_p)]^{1/2} \tag{4.55}$$

The open-circuit voltage is given simply by the product of the resistance and J_{sc}, so that the P.E.M. open-circuit field

$$E_{oc} = (qB/tn_i)(D\tau)^{1/2} \tag{4.56}$$

Equation (4.55) shows that for small signal conditions the short-circuit current increases linearly with light intensity and with magnetic field, and depends on a weighted average mobility to the three-halves power.

A significant feature of equation (4.55) is the dependence of J_{sc} on $\tau^{1/2}$ in contrast with the usual linear dependence of photo-conductance on τ. For this reason measurement of the P.E.M. effect is potentially capable of determining very short lifetimes. In many crystalline samples of PbS for example, where the photo-conductivity is too small to be detected, the author has made P.E.M. measurements and estimated lifetimes less than 10^{-9} seconds. (See *Figure 4.3*.)

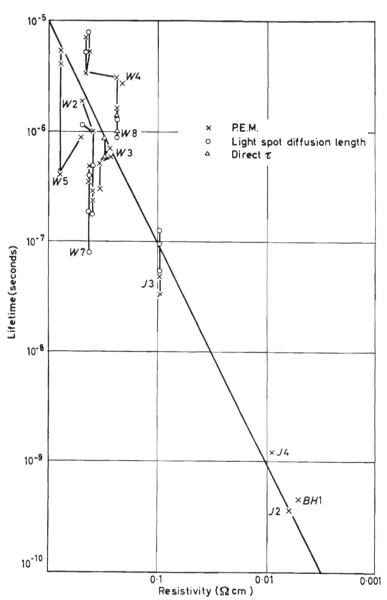

Figure 4.3. *Carrier lifetimes in PbS crystals.* (*Moss*, 1953a; courtesy *Physical Society, London*)

4.4.3 Detailed theoretical treatment

Early theories given by Frenkel (1934, 1935) and Moss et al. (1953) assumed that under open-circuit conditions the net current was everywhere zero. This condition has been shown by van Roosbroeck (1956) to be too stringent, it is only necessary that the *integrated* current across the slab should be zero. The condition that the curl of the electrostatic field must vanish requires a constant P.E.M. field at all depths in the slab and a non-zero curl of local current density. The latter condition implies a circulating current which with the configuration of *Figure 4.2* flows mainly from right to left in the top half of the specimen and from left to right in the lower half of the specimen. Except for end effects, which will be neglected, the currents will flow parallel to the surfaces of the slab, and the direction of maximum carrier concentration gradient will be perpendicular to the illuminated surface.

The treatment is restricted to steady state conditions—studies of the time dependence of the effect have been made by Bulliard (1954b), and Hall (1955). As before, small signal conditions are assumed and we consider a long, wide specimen where end effects are unimportant. Negligible space charge is again assumed, and as in section 4.2.1 we take the approximation of surface generation. No discrimination is made between Hall and conductivity mobilities. It is assumed that there is no net current flow in the direction of irradiation, i.e. $J_y^- + J_y^+ = 0$, although J_x is non-zero. On the basis of these assumptions the current density equations are:—

$$J_x^- = \theta_n J_y^+ + \sigma_n E_x \qquad (4.57)$$

$$J_x^+ = \theta_p J_y^+ + \sigma_p E_x \qquad (4.58)$$

$$-J_y^+ = eD_n \frac{\mathrm{d}p}{\mathrm{d}y} + \theta_n J_x^- + \sigma_n E_y \qquad (4.59)$$

$$J_y^+ = -eD_p \frac{\mathrm{d}p}{\mathrm{d}y} - \theta_p J_x^+ + \sigma_p E_y \qquad (4.60)$$

The continuity equation is

$$-\frac{\mathrm{d}}{\mathrm{d}y} J_y^+ = e\frac{\Delta p}{\tau} \qquad (4.61)$$

From equations (4.57) and (4.58)

$$J_x = \theta J_y^+ + \sigma E_x \qquad (4.62)$$

where $\theta = \theta_p + \theta_n$ is the sum of the Hall angles.

The simplest solution of the problem is obtained for the short-circuit condition when there is no field between the electrodes—i.e. $E_x = 0$.

4.4.4   Short-circuit solution

Eliminating $J_x^- J_x^+$ and E_y we obtain

$$\tau D_n(p_0 + n_0)\frac{d^2 J_y^+}{dy^2} = [p_0 + bn_0 + (p_0 + n_0/b)\theta_n{}^2]J_y^+ \quad (4.63)$$

whose solution is

$$J_y^+ = A \cosh y/f + C \sinh y/f \quad (4.64)$$

where f is an effective diffusion length, given by

$$f^2 = L^2 \left[1 + \theta_n\theta_p \frac{bp_0 + n_0}{bn_0 + p_0}\right]^{-1} \quad (4.65)$$

where $L = (D\tau)^{1/2}$ is the ambipolar diffusion length. The boundary conditions are,

$$y = 0, \quad J_y^+ = eq - es_1\Delta_p = eq + \tau s_1 dJ_y^+/dy \quad (4.66)$$

$$y = t, \quad J_y^+ = es_2\Delta_p = -\tau s_2 dJ_y^+/dy \quad (4.67)$$

Substituting these conditions in equation (4.64) gives the values of A and C, and yields

$$J_y^+ = eq\,\frac{\sinh (t - y)/f + \alpha_2 \cosh (t - y)/f}{(1 + \alpha_1\alpha_2) \sinh t/f + (\alpha_1 + \alpha_2) \cosh t/f} \quad (4.68)$$

where $\alpha_1 = \tau s_1/f$, $\alpha_2 = \tau s_2/f$.

This expression becomes identical with that given by Moss, Pincherle and Woodward (1953) when $\alpha_1 = \alpha_2$ and the magnetic field is so small that θ^2 terms can be ignored, i.e., $f = L$.

The total short-circuit current per unit width of sample is given (from equation 4.62) by

$$J_{sc} = \int_0^t J_x\,dy = \int_0^t \theta J_y^+\,dy$$

$$= \frac{\theta feq(\alpha_2 \sinh t/f + \cosh t/f - 1)}{(1 + \alpha_1\alpha_2) \sinh t/f + (\alpha_1 + \alpha_2) \cosh t/f} \quad (4.69)$$

This expression reduces to that given by van Roosbroeck (1956) for small magnetic fields when we may take $f = L$. For the case of a specimen several diffusion lengths thick so that $e^{t/L} \gg 1$ the short-circuit current becomes

$$J_{sc} = -\theta e f (1 + \alpha_1)^{-1} \qquad (4.70)$$

and

$$J_{sc} = -\theta e q L (1 + \alpha_1)^{-1} \qquad (4.71)$$

for small magnetic fields.

Expression (4.70) has been given by Kurnick and Zitter (1956a) and (4.71) has been given by Moss (1953a). For $\alpha_1 \ll 1$, i.e. negligible surface recombination, (4.71) reduces to the approximate solution given by 4.55. Even if the mobility is not known, this expression can be used to find values for the product of (mobility)^3 and lifetime, as has been done for CdS by Sommers et al. (1956). It may be noted that for

$$\begin{aligned} n_0 \gg p_0, \quad f^2 &= L^2(1 + \theta_p{}^2)^{-1} \\ p_0 \gg n_0, \quad f^2 &= L^2(1 + \theta_n{}^2)^{-1} \\ p_0 = n_0 = n_i, \quad f^2 &= L^2(1 + \theta_p\theta_n)^{-1} \end{aligned} \right\} \qquad (4.72)$$

4.4.5 Photoconductance

The P.E.M. current given by equation (4.66) can be expressed in a simple form in terms of the photoconductance. This latter will be given in the case of no magnetic field by

$$\begin{aligned} \Delta G &= e(\mu_p + \mu_n) \int_0^t \Delta p \, dy \\ &= \frac{e q \tau (\mu_p + \mu_n)(\alpha_2 \cosh t/L + \sinh t/L - \alpha_2)}{(1 + \alpha_1\alpha_2) \sinh t/L + (\alpha_1 + \alpha_2) \cosh t/L} \end{aligned} \qquad (4.73)$$

Hence for small magnetic fields

$$\begin{aligned} J_{sc}/\Delta G &= \frac{B(D/\tau)^{1/2} (\alpha_2 + \tanh t/2L)}{1 + \alpha_2 \tanh t/2L} \\ &= B(D/\tau)^{1/2} \tanh (t/2L + \xi) \end{aligned} \qquad (4.74)$$

where $\tanh \xi = \alpha_2$.

Note that this expression is independent of q, the intensity of irradiation, and also of the recombination velocity of the illuminated surface. For a thick sample where $\tanh t/2L \doteqdot 1$ the relation becomes particularly simple

$$J_{sc}/\Delta G = B(D/\tau)^{1/2} \qquad (4.75)$$

This equation provides a useful method of determining lifetime,

69

assuming D to be known. Values obtained for PbS by this method are shown in *Figure 4.2.*

4.4.6 *Open-circuit condition*

The requirement $\int_0^t J_x \, dy = 0$, in the case of small B, leads by equation (4.62) to

$$J_{sc} = -E_{oc} \int_0^t \sigma \, dy \qquad (4.76)$$

As we are dealing with small signal conditions, σ varies negligibly with y and hence

$$\left. \begin{aligned} J_{sc} &= -\sigma_0 t E_{oc} \\ E_{oc} &= -J_{sc}/\sigma_0 t \end{aligned} \right\} \qquad (4.77)$$

or

It should be noted that J_x is not everywhere zero but is given from equations (4.62) and (4.77) by $J_x = -J_{sc}/t + \theta J_y^+$. For a thick specimen where $\exp t/L \gg 1$ this approximates to

$$J_x = \frac{\theta e q}{1 + \alpha_1} \left[\exp \left(-y/L \right) - L/t \right] \qquad (4.78)$$

This expression shows that current flows along the x axis in one direction near the upper surface and in the opposite direction near the back surface. The change over occurs at $y = L \log L/t$, i.e. a few diffusion lengths below the illuminated surface.

The total circulating current per unit width of sample is

$$J_c = \int_0^{L \log L/t} J_x \, dy \qquad (4.79)$$

which may be seen to be approximately equal to J_{sc}.

4.4.7 *Spectral distribution of sensitivity*

In order to calculate the wavelength variation of the P.E.M. effect it is necessary to consider the generation of hole-electron pairs throughout the material. The rate of generation is given by equation (4.12) and (4.15) replaces (4.61). The boundary conditions are

$$y = 0, \quad -es_1 \Delta p = J_y^+; \qquad y = t, \quad +es_2 \Delta p = J_y^+ \qquad (4.80)$$

For small magnetic fields (where θ^2 terms are neglected and we take $f = L$) equations (4.57)–(4.60) and (4.15) give for the short-circuit condition

$$L^2 \frac{d^2 J_y^+}{dy^2} - J_y^+ = -L^2 eqK^2 e^{-Ky} \qquad (4.81)$$

for which the solution is

$$J_y^+ = A \cosh y/L + C \sinh y/L + \frac{K^2 L^2 eq}{1 - K^2 L^2} \exp(-Ky)$$

Values of A and C may be found from equation (4.80), and the resulting equation for the short-circuit current J_{sc} may be written

$$J_{sc}(1 - k^2)/kLeq\theta =$$

$$\frac{(\alpha_1 + k)(1 - Y \cosh T) + (\alpha_2 - k)(\cosh T - Y)}{+ (1 - k\alpha_2 - Y - k\alpha_1 Y) \sinh T}$$
$$\frac{}{(\alpha_1 + \alpha_2) \cosh T + (1 + \alpha_1\alpha_2) \sinh T}$$

$$(4.82)$$

where $k = KL$, $T = t/L$, and $Y = \exp(-Kt)$.

Equations which reduce to equation (4.82) have been given by Gartner (1957) who has plotted series of curves for the expected spectral behaviour for various values of the parameters α_1, α_2 and T. For negligible surface recombination ($s_1 = s_2 = 0$) this equation becomes:—

$$J_{sc} = \frac{KL^2 eq\theta}{1 - K^2 L^2} \frac{(1 - e^{-Kt}) \sinh t/L + KL(1 + e^{-Kt})(1 - \cosh t/L)}{\sinh t/L}$$

$$(4.83)$$

This expression may be shown to be finite at the value $KL = 1$ where it has the value $J_{sc} = \frac{1}{2}eq\theta L$ for the limiting case of a thick specimen.

At short wavelengths as $K \to \infty$, we find the current rises to a saturation value given by: $(J_{sc})_\infty = eq\theta L$ for the same approximations.

Hence the point on the spectral response curve where the short-circuit current is half its short wavelength value (i.e. the $\lambda_{1/2}$ point) occurs at $KL = 1$. Detailed analysis shows that for any values of surface recombination velocities, i.e. any values of α_1 and α_2, the thick specimen approximation gives $(J_{sc})_\infty = eq\theta L/(1 + \alpha_1)$ and shows that at $KL = 1$ the value of J_{sc} is exactly half this, so that the $\lambda_{1/2}$ point always occurs at $KL = 1$.

Determination of the $\lambda_{1/2}$ point is a useful method of finding the activation energy of a photoconductor, particularly for low

resistance materials of short lifetime where PC effects are in-
conveniently small. Such a case is crystalline PbS, for which the
spectral P.E.M. curve is shown in *Figure 4.4*. Prior to these
measurements (Moss, 1953*a*) it had not been possible to observe
bulk photo-effects in PbS.

Provided the approximations which need to be made to obtain

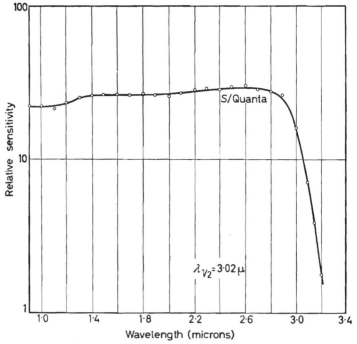

Figure 4.4. *Spectral sensitivity of P.E.M. effect in* PbS.

the relation $KL = 1$ are applicable, measurement of the spectral
distribution of the P.E.M. effect is a suitable method for the
determination of L when adequate data on K are available. It
has the same advantages as measuring the spectral distribution of
the photo-voltaic effect at a probe on the illuminated surface
(see section 4.2.2). However slight differences in spectral sensitivity curves for the P.E.M. effect produced by different surface
treatments have been detected by Kurnick and Zitter (1956*a*).

4.4.8 Behaviour at high magnetic fields

As the magnetic field is increased to the level where Hall angles

72

become appreciable compared with unity, the P.E.M. effect ceases to be linear with the magnetic field. For a high mobility material like Ge, *Figure 4.2* shows that marked non-linearity occurs. For PbS, however, where the mobilities are low, there is no obvious deviation from linearity up to 12,000 G.

Equation (4.70) shows that the form of the relation depends on the surface recombination velocity. If this is low (i.e. $\alpha_1 \ll 1$) then the magnetic field dependence is given by

$$J_{sc} \propto \theta f \propto B/(1 + \mu_f^2 B^2)^{1/2} \qquad (4.84)$$

where the effective mobility μ_f depends on the specimen type

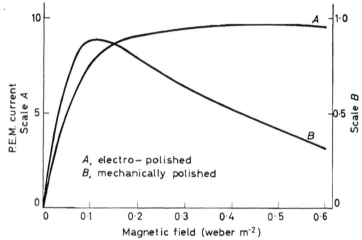

Figure 4.5. Short-circuit P.E.M. current in InSb. (Kurnick and Zitter, 1956b; courtesy John Wiley, New York)

as shown by equation (4.72). Thus the current increases monotonically to a saturation value. For high surface recombination ($\alpha_1 \gg 1$) then

$$J_{sc} \propto \theta f/\alpha_1 \propto \theta f^2 \propto B/(1 + \mu_f^2 B^2) \qquad (4.85)$$

and the current rises to a maximum and falls inversely as B for very high B.

Both these types of behaviour have been observed experimentally by Kurnick and Zitter (1956b). The results obtained are shown in *Figure 4.5*.

Measurements of the saturation of the P.E.M. effect at high magnetic fields have been reported by de Carvalho (1956).

73

Photoelectro-magnetic effects in non-uniform magnetic fields have been treated experimentally and theoretically by Tauc (1956).

4.4.9 Behaviour under strong irradiation

In the large signal case the rate of optical production of hole-electron pairs is assumed to be so large that the hole and electron densities are markedly increased and may be considered equal i.e. a pseudo-intrinsic condition exists. Assume that s_1 and s_2 remain unaltered at their equilibrium values and replace D by D_i. Depending on the actual recombination processes occurring τ may or may not show concentration dependence:

(1) Constant lifetime assumed, independent of carrier concentrations

Equation (4.61) remains linear, and for small fields J_{sc} is given by (4.69) with $f = L$. Also $E_x = -J_{sc}/G$ where the conductance $G \sim \Delta G \gg G_0$. Hence

$$E_x \doteqdot B(D/\tau)^{1/2} \frac{\tanh t/2L + \alpha_2}{1 + \alpha_2 \tanh t/2L} \tag{4.86}$$

This expression is independent of q, indicating that the P.E.M. field saturates at sufficiently high values of irradiation intensity.

For thick specimens where $\tanh t/2L \doteqdot 1$,

$$E_x = B(D/\tau)^{1/2} \tag{4.87}$$

For thin specimens where $\tanh t/2L \sim 0$,

$$E_x = B(D/\tau)^{1/2}\alpha_2 = Bs_2 \tag{4.88}$$

This latter equation, which was originally given by Moss, Pincherle and Woodward (1953) as applying on the back surface, is here shown to hold on both surfaces. It forms a particularly simple way of measuring s_2 the recombination velocity of the back surface, when it is applicable. As specified here, the thickness of the slab must be small compared with a diffusion length—which is equivalent to the proviso that bulk recombination can be neglected. Experimental saturation curves often show slight difference between voltages on the front and back surfaces, presumably due to differences between s_1 and s_2.

The case of arbitrary illuminating intensity has been treated by van Roosbroeck (1956).

(2) Other Recombination Laws

The effects of bimolecular recombination laws, either in the bulk material or at the surface, have been considered by Pincherle (1956). For the experimental arrangement o *Figure 4.2*, where the ends of the specimen are covered by the electrodes, Pincherle's results for the open-circuit voltage under very strong irradiation may be summarized as follows:—

(*a*)   Linear bulk, linear surface, recombination. Saturation occurs.

$$E_{oc} = -B(D/\tau)^{1/2} \text{ for thick specimens}$$

$$= -Bs_2 \text{ for thin specimens}$$

(*b*)   Linear bulk, quadratic surface, recombination. Saturation occurs.

$$E_{oc} = -B(D/\tau)^{1/2} \text{ for thick specimens}$$

$$= -2BD/t \text{ for thin specimens}$$

(*c*)   Quadratic bulk, linear surface recombination. No saturation.

$$E_{oc} \propto q^{1/3}$$

(*d*)   Both quadratic. No saturation.

$$E_{oc} \propto q^{1/4}$$

4.5   Photomechanical effects, or photomotive forces

As shown in section 4.4.6, as a consequence of the P.E.M. effect a considerable circulating current may be set up in an open-circuit photoconductor. By interaction with a magnetic field such circulating currents can obviously give mechanical effects.

In the configuration considered in section 4.4, i.e. a rectangular parallelopiped of photoconductor with the illumination and magnetic field mutually perpendicular and along the axes of the specimen, this circulating current lies completely in a plane perpendicular to the magnetic field and so no net mechanical force is produced by the interaction if the field is homogeneous. However suppose the magnetic field is tilted (in the *yz* plane of *Figure 4.2*) so that it makes an angle ϕ with the Oz axis, then only the component of B perpendicular to the diffusion gradient, i.e. $B \cos \phi$, will be effective in producing separation of the hole-electron pairs, and consequently the P.E.M. currents will be proportional to $B \cos \phi$ in lieu of B.

The circulating current will still lie in the xy plane and it will now interact with the component of the magnetic field in this plane, i.e. with $B \sin \phi$, to give a torque which is consequently proportional to $B^2 \sin \phi \cos \phi$.

Consider a circular cylinder, radius r, with its axis vertical, placed in a horizontal magnetic field with irradiation in the horizontal plane making an angle of ϕ with that of the magnetic field. Let the illumination cover a narrow vertical strip of width w over the total length of the cylinder X. Assuming strongly absorbed radiation and a thick specimen the total vertical current will be given from equation (4.78) as

$$wXJ_x = wX \left(\frac{\theta eq \cos \phi}{1 + \alpha_1} \right) (\exp - y/L - L/2r) \qquad (4.89)$$

The tangential force on this current will be

$$F = wXB\theta eq \cos \phi \sin \phi \ (\exp - y/L - L/2r)/(1 + \alpha_1) \qquad (4.90)$$

and the total resulting torque

$$T = \int_0^{2r} wX(r - y) \left(\frac{B\theta eq \cos \phi \sin \phi}{1 + \alpha_1} \right) (\exp - y/L - L/2r) \ \mathrm{d}y \qquad (4.91)$$

$$= wX \left(\frac{LrB\theta eq \cos \phi \sin \phi}{1 + \alpha_1} \right) \text{ if } L \ll r \qquad (4.92)$$

$$= wX \left(\frac{LrB^2 eq(\mu_n + \mu_p)}{2(1 + \alpha_1)} \right) \text{ for } \phi = 45° \qquad (4.93)$$

$$= \frac{QLrB^2 e}{2(1 + \alpha_1)} (\mu_n + \mu_p) \text{ where } Q = qwX \text{ is the total irradiation.}$$

A similar expression to the above has been given by Garreta and Grosvalet (1956), following the original theory of Aigrain and Garreta (1954). The existence of the effect has been observed in germanium by Gousseland (see Garreta and Grosvalet, 1956) who verified that the magnitude of the torque increased linearly with the light intensity and with the square of the magnetic field.

Measurement of this photomotive force has the unique advantage that no electrical contacts need to be made to the specimen.

4.6 PHOTO-EFFECTS IN p–n JUNCTIONS

4.6.1 *Utilization of junction photo-effects*

p–n Junctions may be used as photocells in several different ways. They have the attraction that they can be used as voltage (and power) generators without the need for a polarizing voltage. If they *are* used with a polarizing voltage in the manner of photo-conductive detectors they have the advantage of considerably higher resistance and responsivity than a comparable slab of homogeneous semi-conductor.

Normally *p-n* junctions will be used with radiation of wavelengths not much less than the absorption edge—i.e. in the near i.r. and the visible. However Pfister (1956) has shown that they may be used for measurement of x-ray intensities and that the e.m.f. produced increases linearly at low intensities.

p-n Junction cells can be used with the irradiation either in the plane of the junction or normal to the plane of the junction.

4.6.2   *Illumination parallel to the junction*

Consider a *p-n* junction of width w and thickness t, illuminated in the direction of t in a very narrow strip r on either side of the junction so that q electron-hole pairs are generated per second per unit area. The total rate of quanta generation is thus $2rwq$. Consider the short-circuit case where electrodes on the ends of the *p* and *n* regions are connected together. Of each electron-hole pair generated on the *n*-type side of the junction the hole will be attracted to the junction and contribute to the photocurrent. Similarly the electrons on the *p*-type side will give photocurrent.

Assuming that there is no significant potential drop in the system each of these carriers will contribute a charge e to the external circuit—the effects of holes and electrons adding as they are moving in opposite directions. The total short circuit current will thus be

$$J_{sc} = 2rwqe \qquad (4.94)$$

If the specimen is uniformly illuminated over large areas on either side of the junction, then only carriers generated within a diffusion length of the junction will reach the junction and give a photocurrent. In this case the total photocurrent is

$$J_{sc} = wqe(L_e + L_h) \qquad (4.95)$$

and the current density is

$$J_{sc}/wt = qe(L_e + L_h)/t \qquad (4.96)$$

where L_h is the diffusion length of holes in the *n* region and L_e is the diffusion length of electrons in the *p* region.

In the open-circuit case the flow of carriers starts as indicated in equation (4.95) but then a voltage builds up between the electrodes causing a reverse current flow across the junction. In equilibrium these two currents are equal. Now the current density flowing across a p–n junction under the influence of an applied voltage V is given by Shockley (1950) as

$$I = (kT/e)(\exp eV/kT - 1)(b\sigma_i^2/[1 + b]^2)(1/\sigma_p L_e + 1/\sigma_n L_h) \quad (4.97)$$

$$= e(n_0 D_e/L_e + p_0 D_h/L_h)(\exp eV/kT - 1) \quad (4.98)$$

where n_0 is the equilibrium concentration of the electrons in the p-region.

The equilibrium open-circuit photo-voltage is obtained by equating the current densities given by equations (4.96) and (4.98) which gives

$$V_{oc} = (kT/e) \log \left[\frac{q(L_e + L_h)/t}{n_0 D_e L_e + p_0 D_h/L_h} + 1 \right] \quad (4.99)$$

If the mobility ratio is not far from unity so that we can make the simplifying approximations $D_e = D_h$ and $L_e = L_h$, and if furthermore we take $n_0 = p_0$ (i.e. the two sides are equally doped) then

$$V_{oc} = (kT/e) \log (1 + q\tau/n_0 t) \quad (4.100)$$

4.6.3 Illumination perpendicular to the junction

Consider a junction which consists of a very thin p-layer extending from $y = 0$ to the junction plane at $y = t$, with a relatively thick n-layer extending to $y = Y$. The illumination falls normally on the $y = 0$ surface of the p-layer and short wavelength radiation (i.e. surface generation) is assumed.

Ignoring the field component of the current in comparison with the diffusion current, we have for electrons in the p-region at equilibrium,

$$D_e(d^2 \Delta n/dy^2) = \Delta n/\tau_e \quad (4.101)$$

with the boundary conditions

$$y = 0, \quad -D_e \, d\Delta n/dy = q - s\Delta n; \quad y = t, \quad n = n_0 \exp eV/kT \quad (4.102)$$

These equations give for the electron current density at the junction

$$J^- = \frac{eD_e}{L_e} \left[\frac{n_0(\exp eV/kT - 1)(\alpha_1 \cosh t/L_e + \sinh t/L_e) - qL_e/D_e}{\alpha_1 \sinh t/L_e + \cosh t/L_e} \right] \quad (4.103)$$

We assume that the p-layer is so thin that $t \ll L_e$, when we put $\cosh t/L_e = 1$, and $\sinh t/L_e = t/L_e$. In order for surface recombination to be unimportant on the denominator it is necessary that

$$\alpha_1 t/L_e \ll 1 \qquad \text{or} \qquad s \ll D_e/t \qquad (4.104)$$

It should be noted that this is a less stringent requirement than the usual one that $\alpha_1 \ll 1$ or $s \ll D_e/L_e$, since $t \ll L_e$. For the α_1 term to be negligible in the numerator it is necessary that

$$\alpha_1 \ll t/L_e \qquad \text{or} \qquad s \ll t/\tau_e \qquad (4.105)$$

which is a much more difficult requirement to meet.

If these approximations are valid we obtain from (4.103)

$$J^- = e n_0 t (\exp eV/kT - 1)/\tau_e - eq \qquad (4.106)$$

In the n-region we have for the holes,

$$D_h \, \mathrm{d}^2 \Delta p / \mathrm{d}y^2 = \Delta p / \tau_h \qquad (4.107)$$

with the boundary conditions

$$y = t, \quad p = p_0 \exp eV/kT \qquad y = t + Y, \quad p = p_0 \qquad (4.108)$$

The solution of the above equations is

$$J^+ = (e p_0 D_h / L_h)(\exp eV/kT - 1) \coth Y/L_h \qquad (4.109)$$

If $Y \gg L_h$ so that we may put $\coth Y/L_h = 1$ this becomes

$$J^+ = e p_0 D_h (\exp eV/kT - 1)/L_h \qquad (4.110)$$

and the total current density across the junction is

$$J = e(\exp eV/kT - 1)(n_0 t/\tau_e + p_0 D_h/L_h) - eq \qquad (4.111)$$

The short-circuit photocurrent is simply $J_{sc} = -eq$ and the open-circuit photo-voltage is evidently given by putting $J = 0$, i.e.

$$V_{oc} = (kT/e) \log [q/(n_0 t/\tau_e + p_0 D_h/L_h) + 1] \qquad (4.112)$$

4.6.4 The solar battery

With the possibility of fabricating large area p–n junctions of the configuration treated in section 4.6.3, in silicon for example, it has become feasible to generate appreciable quantities of electrical power from solar radiation. As we are now interested in *power* from the illuminated junction neither the short-circuit current nor open-circuit voltage equations apply, but we must treat the case of the junction feeding a resistive load Z, which is matched so that the power W delivered to it is a maximum. This latter condition is expressed by

$$dW/dZ = d(J^2Z)/dZ = 0$$

From the latter

$$2dJ/J = -dZ/Z$$

so that

$$dW/dJ = 0 \qquad (4.113)$$

Writing j for J/e and putting

$$j_0 = n_0t/\tau_e + p_0D_h/L_h = n_0t/\tau_e + p_0L_h/\tau_h \qquad (4.114)$$

we have from (4.111)

$$j + q = j_0(\exp eV/kT - 1) \qquad (4.115)$$

Putting

$$eV/kT = ajZ \qquad (4.116)$$

where a is a factor proportional to the cross-sectional area (A) of the junction, we have from equation (4.115)

$$ajZ = \log [1 + (j + q)/j_0] \qquad (4.117)$$

so that the power W, which is proportional to j^2Z, is proportional to $j \log (j_0 + j + q)/j_0$. Hence from equation (4.113)

$$\frac{d}{dj} [j \log (j_0 + j + q)/j_0] = 0 \qquad (4.118)$$

or

$$\frac{j_0 + j + q}{j_0} \log \frac{j_0 + j + q}{j_0} = -j/j_0 \qquad (4.119)$$

This equation can be solved for any given values of j_0 and q to find the optimum value of j and hence of Z and W.

For strong illumination such that $q \gg j_0$ the solution of (4.119) is

$$-j/(j + q) = \log q/j_0 - \log \log q/j_0 \qquad (4.120)$$

to a good degree of approximation, from which it may be seen that $j = -q$ within a few per cent—i.e. the current under matched conditions approximates to the saturation current $J_{sc} = -eq$. Also as $\exp eV/kT \gg 1$ the voltage is given by

$$\exp eV/kT = (j + q)/j_0 = q/j_0(1 + \log q/j_0 - \log \log q/j_0)$$

or

$$eV/kT \doteqdot \log q/j_0 - \log \log q/j_0 \qquad (4.121)$$

The power developed in the load per unit area of junction is

$$W = ejV = qkT \,(\log q/j_0 - \log \log q/j_0) \qquad (4.122)$$

From this equation it may be seen that the solar battery with its matched load operates as though each photon provides $\sim kT \log q/j_0$ joules of energy to the external circuit. Clearly for maximum output it is necessary to make q as large, and j_0 as small as possible.

For the latter, which is given by equation (4.114), we may assume that $n_0 \sim p_0$—i.e. both sides of the junction are equally doped. It is not difficult to make t small enough for the second term to be the dominant one, and we therefore assume $j_0 = p_0 L_h/\tau_h$. Thus p_0 and correspondingly n_0, must be kept as small as possible, which means that the two sides of the junction must be heavily doped, since if n_d is the density of ionized donors, $p_0 = n_i^2/n_d$.

Assuming that the total incident solar radiation is fixed, q can only be varied by changing the cut-off wavelength of the photoconductor, i.e. by varying the percentage of the solar radiation whose quanta are large enough to produce photo-effects. Thus q increases steadily as the cut-off wavelength increases. However increasing the cut-off wavelength and hence reducing the activation energy increases j_0 because decrease in activation energy raises the intrinsic carrier concentration. To obtain an estimation of this optimum energy gap we make the broad simplification that for any photoconductor used all parameters would be the same except for the energy gap E_g. Taking $n_d = 10^{18}$ and $L_h/\tau_h = 10^3$ so that

$$j_0 \doteqdot 10^3 p_0 \doteqdot 10^{-15} n_i^2 \doteqdot 10^{+23} \exp -E_g/kT \qquad (4.123)$$

and taking the solar radiation to be black-body radiation of $6000°\mathrm{K}$, with the intensity $0 \cdot 1$ watt/cm², we can calculate the efficiency. This is given by

$$\eta = kT(\log q/j_0 - \log \log q/j_0)q/1 \cdot 4q_0 \qquad (4.124)$$

where q_0 is the total incident photon intensity, which from the above parameters is $4 \cdot 5 \times 10^{17}$ quanta/cm²/sec. The factor $1 \cdot 4$ is the *average* value of a solar photon in eV. Using values of q/q_0 obtained directly from a radiation slide-rule the above expression, which ignores reflection losses, is plotted in *Figure 4.6*.

It will be seen that curve A peaks at $1 \cdot 4$ eV at 29 per cent efficiency. Curve B is plotted on the assumption that effective absorption takes place only up to a wavelength 10 per cent lower than that equivalent to the activation energy. This reduces the optimum activation energy to $1 \cdot 3$ eV, with a peak efficiency which is 25 per cent. These optimum activation energies are somewhat lower than those found by Rittner (1956) although the efficiencies are about the same as those calculated by Rittner (1956) and Cummerlow (1954).

Loferski (1956) has shown that atmospheric absorption bands influence the value of the optimum activation energy. Jenny *et al.* (1956) have made solar energy converters from GaAs which has an energy cut-off at 1·35 eV. Considering that these units were not

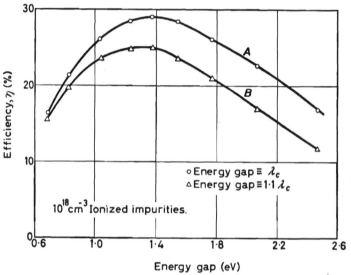

Figure 4.6. Efficiency of solar battery as function of photoconductor energy gap

of optimum design, the efficiencies measured by these workers show that this material is very promising, and that in fact this energy gap may prove to be about the best.

The practical design of solar battery elements has been discussed by Chapin *et al.* (1954), Prince (1955) and Prince and Wolf (1958). Working units with efficiencies up to 14 per cent have been made.

Analogous to the solar battery is the 'atomic battery' where power is produced by irradiating a silicon p–n junction from a radioactive source (Pfann and van Roosbroeck, 1954). Using β particles from a strontium-90 source Rappaport (1954) has obtained 1 μW of electrical power.

CHAPTER 5

MAGNETO OPTICAL EFFECTS

5.1 FARADAY EFFECT

5.1.1 Introduction

IF a plane polarized beam of radiation is passed through a block of optically inactive material, it is found that the plane of polarization is rotated when a magnetic field is applied along the direction travelled by the radiation. On reversing the magnetic field the direction of rotation is reversed. Similarly reversing the direction of the beam of radiation reverses the angle of rotation, and consequently reflecting a beam to and fro through the specimen progressively increases the total rotation.

The general explanation of this effect is that the plane wave in the medium can be considered to be split up into two oppositely polarized circular vibrations which are propagated with different velocities in the presence of the field. One wave travels faster than, and the other slower than in zero magnetic field.

5.1.2 Faraday effect in dielectrics

We shall first treat the effect on the basis of the vibration of bound electrons in a dielectric as used for dispersion theory in Chapter 2.

Consider the case of a magnetic induction B applied in the z direction, in which the wave is travelling. Using the notation of equation (2.1), but making the simplifying assumption that the damping term can be neglected (which will be valid at frequencies well removed from the resonance frequency) we obtain

$$\left.\begin{aligned}
\frac{d^2x}{dt^2} + \omega_0^2 x + \frac{Be}{m^*}\frac{dy}{dt} &= \frac{e}{m^*}E_x e^{i\omega t} \\
\frac{d^2y}{dt^2} + \omega_0^2 y - \frac{Be}{m^*}\frac{dx}{dt} &= \frac{e}{m^*}E_y e^{i\omega t}
\end{aligned}\right\} \tag{5.1}$$

where E_x and E_y are the impressed fields in the x and y directions respectively. Multiplying the second equation by i and adding we obtain

$$\frac{d^2(x+iy)}{dt^2} + \omega_0^2(x+iy) - \frac{iBe}{m^*}\frac{d(x+iy)}{dt} = \frac{e}{m^*}(E_x + iE_y)e^{i\omega t} \tag{5.2}$$

This equation shows that the electron vibrates with a complex amplitude given by

$$x + iy = \frac{(E_x + iE_y)e/m^*}{\omega_0{}^2 - \omega^2 + Be\omega/m^*}$$ (5.3)

so that by analogy with equation (2.3),

$$(n_1 - ik_1)^2 = 1 + (Ne^2/m^*\varepsilon_0)/(\omega_0{}^2 - \omega^2 + Be\omega/m^*)$$ (5.4)

As we have assumed no damping the right hand side of this expression is of course real and k_1 is zero. Hence

$$n_1{}^2 = 1 + \frac{Ne^2/m^*\varepsilon_0}{\omega_0{}^2 - \omega^2 + B\omega e/m^*} = 1 + \frac{Ne^2/m^*\varepsilon_0}{\omega_0{}^2 - \omega^2 + \omega\omega_c}$$ (5.5a)

where

$$\omega_c = Be/m^*$$ (5.6)

is the cyclotron angular frequency and m^* is the effective mass of the charge carrier. For the wave which is circularly polarized in the opposite sense the sign of the ω_L term is reversed giving

$$n_2{}^2 = 1 + (Ne^2/m^*\varepsilon_0)/(\omega_0{}^2 - \omega^2 - \omega\omega_c)$$ (5.5b)

As $\omega_c \ll \omega$, these expressions may be rewritten

$$n_1{}^2 = 1 + (Ne^2/m^*\varepsilon_0)/(\omega_0{}^2 - (\omega + \tfrac{1}{2}\omega_c)^2)$$ (5.5c)

$$n_2{}^2 = 1 + (Ne^2/m^*\varepsilon_0)/(\omega_0{}^2 - (\omega - \tfrac{1}{2}\omega_c)^2)$$ (5.5d)

The effect of the magnetic field on the refractive index is thus to change it to the values it would have if the frequency were changed by $\pm\tfrac{1}{2}\omega_c$, i.e. by the Larmor precession frequency.

The rotation of the plane of polarization in radians per unit length is given by

$$\theta = \omega(n_2 - n_1)/2c$$ (5.7)

As n_1 and n_2 differ only slightly we may put

$$n_2 - n_1 = (n_2{}^2 - n_1{}^2)/2n$$

so that from equation (5.5) we obtain

$$\theta = Ne^3B\omega^2/2cnm^{*2}\varepsilon_0(\omega_0{}^2 - \omega^2)^2$$ (5.8)

if we again neglect $\omega_c{}^2$ compared with ω^2.

This expression may be seen to be related simply to the dispersion of the material in the absence of the magnetic field, since then

$$n^2 = 1 + (Ne^2/m^*\varepsilon_0)/(\omega_0{}^2 - \omega^2)$$ (5.9)

if we again neglect the damping term. Hence

$$\theta = (\omega\omega_c/2c)(dn/d\omega) \qquad (5.10)$$

It is clear that by measuring θ in a given magnetic field and assuming the dispersion curve to be known, we may obtain values of ω_c and hence of the effective mass of the charge carriers from equation (5.6). This is done for example in section 15.5.

5.1.3 Faraday effect and free carriers

In the case of free carrier absorption, we put $\omega_0 = 0$, but include the damping term in the equation for the motion of the charged carriers. The equations for the deflection of a free electron by the two circularly polarized waves into which the plane polarized wave may be resolved then are:

$$m\frac{d^2s}{dt^2} + gm\frac{ds}{dt} - iBe\frac{ds}{dt} = eEe^{\pm i\omega t} \qquad (5.11)$$

where $s = x \pm iy$ is the complex deflection and g is the angular damping frequency or reciprocal relaxation time.

The only useful solutions of this equation will be of the form $s = Ke^{\pm i\omega t}$, where $-\omega^2 K \pm i\omega gK \pm (Be\omega/m)K = eE/m$.

Hence

$$K = (Ee/m\omega)/(-\omega \pm Be/m \pm ig)$$

which is the complex amplitude of the deflection.

Now the complex dielectric constant is given by

$$(n - ik)^2 = \varepsilon + (Ne/\varepsilon_0)(s/Ee^{\pm i\omega t})$$
$$= \varepsilon + (Ne^2/m\omega\varepsilon_0)/(-\omega \pm Be/m \pm ig) \qquad (5.12)$$

Thus

$$n^2 - k^2 = \varepsilon - \frac{(Ne^2/m\omega\varepsilon_0)(\omega \pm Be/m)}{(\omega \pm Be/m)^2 + g^2} \qquad (5.13a)$$

and

$$2nk = (Ne^2g/m\omega\varepsilon_0)/([\omega \pm Be/m]^2 + g^2) \qquad (5.13b)$$

Assume the absorption is small and proportional to (wavelength)^2 so that $k^2 \ll n^2$ and $g^2 \ll \omega^2$ and that the magnetic field is so small that $(Be/m)^2 \ll \omega^2$, then

$$n_1^2 - n_2^2 = (Ne^2/m\varepsilon_0)(2Be/m\omega^3) \qquad (5.14)$$

Now

$$\theta = \omega(n_1 - n_2)/2c = \omega(n_1^2 - n_2^2)/4nc$$

therefore

$$\theta = BNe^3/2nc\varepsilon_0 m^2\omega^2 \qquad (5.15)$$

85

To the same degree of approximation as used above,

$$n^2 = \varepsilon - Ne^2/m\varepsilon_0\omega^2$$

therefore

$$n \, \mathrm{d}n/\mathrm{d}\omega = Ne^2/m\varepsilon_0\omega^3$$

so that

$$\theta = (Be/2mc)\omega \, \mathrm{d}n/\mathrm{d}\omega = \tfrac{1}{2}(\omega_o\lambda/c) \, \mathrm{d}n/\mathrm{d}\lambda \qquad (5.16)$$

This equation is the same as equation (5.10). It has been derived by Mitchell (1955) by an alternative method. Phenomenologically the Faraday rotation is a measure of the dispersion due to cyclotron resonance absorption.

A more detailed treatment by Stephen and Lidiard (1958/9) starting from the Boltzmann transport equation shows that for a degenerate semi-conductor with spherical energy surfaces an equation identical with (5.15) is obtained, provided that the effective mass is interpreted as

$$\frac{1}{m^*} = \frac{1}{\hbar^2}\left(\frac{\mathrm{d}E}{k\mathrm{d}k}\right)_F \qquad \text{or} \qquad m^* = \frac{\hbar^2}{2}\left(\frac{\mathrm{d}k^2}{\mathrm{d}E}\right)_F \qquad (5.17)$$

where the subscript F means the value at the Fermi surface. For parabolic energy bands the above expressions are of course identical with the conventional definition of m^*, but for InSb for example, where the conduction band is non-parabolic, the difference is important.

The most interesting feature of equation (5.15) is that the Faraday rotation depends on the square of the effective mass, so that a measurement of this rotation together with the values of the other parameters—which are not difficult to determine—is a potentially accurate way of finding the effective mass. For 1 cm thickness of high conductivity n type germanium (e.g. $\sim 50\Omega^{-1}$ cm^{-1}) using a field of 10,000 G and a wavelength of 5 μ, the rotation is approximately $0.1/m_e^{*2}$ degrees, i.e. $\sim 1°$. For InSb, using a wavelength of 15 μ and the same carrier density (i.e. $\sim 10^{17}$ cm^{-3}) the rotation for a specimen 1 mm thick would be $\sim 0.1/m_e^{*2}$ degrees. As $m_e^* \sim 0.02$, the rotation in this case would be extremely large and easily measurable. Such measurements are described in section 16.10.

5.1.4 Faraday effect at very long wavelengths

At wavelengths so long that $\omega^2 \ll g^2$ (for example at *cm* wavelengths) the conductivity in the absence of the magnetic field, which is given by $2nk\omega$, will be independent of frequency. The real part of the dielectric constant will still depend on frequency, so that there will be dispersion and hence a Faraday effect.

Making the assumption that the absorption is still so small that $2nk \ll n^2 - k^2$ we obtain from equation (5.13a)

$$n_1{}^2 = \varepsilon - \frac{(Ne^2/\omega m^* \varepsilon_0)(\omega - Be/m^*)}{g^2 + (Be\omega/m^* - \omega)^2} \qquad (5.17)$$

Hence, putting $\omega_c = Be/m^*$,

$$n_2{}^2 - n_1{}^2 = \frac{2(Ne^2/\omega m^* \varepsilon_0)(\omega_c{}^2 - \omega^2 + g^2)\omega_c}{(\omega^2 + \omega_c{}^2 + g^2)^2 - 4\omega^2\omega_c{}^2} \qquad (5.18)$$

Ignoring ω^2 compared with g^2, and assuming that magnetic fields are small so that $g^2 \gg Be\omega/m^*$ we have

$$\theta = \frac{\omega}{2c}(n_2 - n_1) = \frac{\omega}{4nc}(n_2{}^2 - n_1{}^2) = BNe^3/2g^2cnm^{*2}\varepsilon_0 \quad (5.19)$$

Substituting $g = e/\mu m^*$ and $\sigma = Ne\mu$ we obtain

$$\theta = \tfrac{1}{2}\sigma\mu B/cn\varepsilon_0 \text{ radians per unit length} \qquad (5.20)$$

Equation (5.20) has been given by Rau and Caspari (1955) who also give the more complex formulae for use when the absorption term $2nk$ is not negligible.

Equation (5.20) predicts large effects in high mobility semi-conductors. For Ge for example, taking 10^{22} electrons/m³, $\mu = 0.36$ m²/V.sec and $B = 1$ (10,000 G) we find $\theta = 10^4$ radians/m or 10 radians/mm. Even larger values have been estimated by Klinger and Chaban (1956).

The above equations assume an energy independent relaxation time $(\tau = 1/g)$ and an isotropic effective mass. The extension of the theory to ellipsoidal energy surfaces has been discussed by Lax and Roth (1955) and Rau and Caspari (1955). A recent quantum mechanical treatment of the Faraday effect has been given by Stephen (1958a).

5.2 CYCLOTRON RESONANCE

In a cyclotron resonance experiment the current carriers in a solid move in spiral orbits about the axis of the applied magnetic induction B. If a high frequency field is applied at right angles to the static magnetic field, absorption occurs, and reaches a maximum when the applied frequency is equal to the cyclotron resonance frequency.

As can be seen from equation (5.13b) the absorption term is a maximum when

$$\omega_c = Be/m^* \qquad (5.21)$$

Phenomenologically this result can be derived as the condition for a stable orbit, namely that the magnetic force on the carrier just equals the centrifugal force, i.e.

$$Be\omega_c r = m^* r \omega_c^2 \qquad (5.22)$$

In order for the absorption to be observable, it is necessary not only for condition (5.20) to be satisfied, but also that the collision relaxation time τ be long enough for a carrier to execute a considerable fraction of a complete revolution between collisions. This means that $\omega_c \tau$ must be at least about unity. In terms of equation (5.13b) we can see that to get a pronounced resonance it is necessary that

$$\omega_c \gg g \qquad \text{or} \qquad \omega_c \tau \gg 1 \qquad (5.23)$$

Curves are given by Kittel (1956) for $\omega_c \tau = 0.2$, 1.0 and 2.0 from which it may be seen that even with $\omega_c \tau = 1$ the resolution is very poor, and $\omega_c \tau \gg 2$ is preferable.

Measurement of the cyclotron resonance frequency ω_c is the most direct method yet devised of obtaining the effective masses of current carriers. Its use for this purpose was first suggested by Dingle (1952). The sign of the charge carriers can be verified by the use of circularly polarized radiation, since only when the direction of rotation of the charge carriers corresponds with the direction of rotation of the circularly polarized radiation will energy be absorbed at resonance. In addition to the values of the effective mass, an estimate of the relaxation time can generally be made from the width of the resonance line. The detailed study of cyclotron resonance for various crystal directions, for varying wavelengths and for excited states, has provided a most powerful tool for the investigation of energy bands in semi-conductors.

Until recently cyclotron resonance experiments had been carried out only at radio frequencies (i.e. up to 24,000 Mc/s) and the condition $\omega_c \tau \gg 1$ could only be satisfied by increasing τ by cooling to liquid helium temperatures. Under such conditions much excellent work has been carried out on germanium and silicon by Kittel (1956), Kip (1954), Lax et al. (1954), Dexter et al. (1956) and Lax (1955).

An alternative approach to the problem was made by Burstein, Picus and Gebbie (1956) who satisfied the condition by increasing ω_c by working at infra-red wavelengths. They succeeded in measuring an absorption maximum in InSb using a wavelength of 41 μ and magnetic fields up to 60,000 G, from which they found $m_e^* = 0.015m$. It may be noted that their absorption curve was asymmetrical, possibly due to the non-parabolic nature of the

88

conduction band (Wallis, 1958). As $\tau = m^*\mu/e$, a good crystal of InSb for which $\mu_e = 6m^2/V$. sec should give $\omega_c\tau \sim 25$ at the wavelength (41 μ) that was used.

Experiments have since been carried out at wavelengths as low as 10 μ by Keyes et al. (1956) by using pulsed magnetic fields up to 300,000 G. By working at a variety of wavelengths and fields these workers have succeeded in studying the energy dependence of the effective mass, which essentially gives an $E:k$ plot for the conduction band. This same group of workers (Zwerdling et al., 1956) have measured the shift of the absorption edge at the same high magnetic fields.

In the low temperature microwave work there are not in general enough free carriers present to give adequate absorption, and it is necessary to excite some carriers from impurity levels. Initially for germanium this excitation was provided by the R.F. field itself. Later optical excitation was employed and it was found that new resonance peaks appeared. These were attributed to excited states, which therefore became accessible to study. Of particular interest is the fact that use of sufficiently short wavelength radiation enables electrons to be excited into the conduction band at $\tilde{k} = 0$, so that this part of the germanium band structure can be investigated experimentally.

The general quantum-mechanical theory of cyclotron resonance in semi-conductors has been given by Luttinger (1956). The classical theory has been worked out for warped energy surfaces by Zeiger, Lax and Dexter (1957) assuming an energy independent relaxation time. Line shapes and harmonic intensities are calculated and shown to be in reasonable agreement with observations in Ge and Si at low (i.e. radio) frequencies. The classical and quantum theories agree except at very low quantum numbers where they predict slightly different resonance frequencies in particular cases.

The effect of the non-parabolic energy bands in InSb has been treated by Wallis (1958). Dresselhaus et al. (1955a) have shown that in small specimens of InSb the radio frequency cyclotron resonance can be influenced by plasma oscillations.

5.3 OSCILLATORY MAGNETO-OPTIC EFFECTS

According to Burstein and Picus (1957), Zwerdling, Lax and Roth (1957) and Roth et al. (1958) strong magnetic fields cause the quasi-continuous levels in the allowed energy bands to coalesce into magnetic sub-bands, and optical transitions between the various sub-bands may then take place.

For a magnetic field B along the z axis the sub-bands have energy levels given by

$$E_n = \hbar^2 k_z{}^2/2m^* + \hbar\omega_c(n + \tfrac{1}{2})$$

where the quantum number $n = 0, 1, 2 \ldots$ and the energy separation between the sub-bands is given by the cyclotron energy, $\hbar\omega_c = e\hbar B/m^*$. Well defined sub-bands should occur when the relaxation time $\tau > 1/\omega_c$.

Experimentally Burstein and Picus (1957) have observed three absorption peaks in the region of the main absorption edge of InSb using fields $\sim$50,000 G. Zwerdling, Lax and Roth (1957) find four well-defined absorption maxima for InSb. When the photon energies for these maxima are plotted against magnetic field, they can all be extrapolated back to zero field to give the same value of photon energy. This is the true activation energy, the figure obtained being $0\cdot180 \pm 0\cdot002$ eV.

Interpretation of the slopes of the $h\nu - B$ plots is less certain. If the first three maxima are taken as $(n + \tfrac{1}{2})\hbar\omega_c$ where $n = 0$, $1, 2 \ldots$, they all give the same value of effective mass, namely $m^* = 0\cdot024m$. As this value is rather high compared with other determinations, Zwerdling, Lax and Roth (1957) suggest that each sub-band or Landau level may be split into two by spin-orbit interaction. Accordingly they take the average of the first two absorption maxima as corresponding to $n = 0$, and that of the next two as $n = 1$, and obtain effective masses of $m^* = 0\cdot014m$ and $0\cdot013m$.

Zwerdling and Lax (1957) and Zwerdling, Lax and Roth (1957, 1958) have made an extensive study of the germanium absorption edge for direct transitions. They obtain the precise value of $0\cdot803 \pm 0\cdot001$ eV for this energy gap at room temperature, and estimate the effective mass of carriers in this part of the conduction band—i.e. at $\tilde{k} = 0$—to be about $0\cdot036$ free electron masses with slightly larger values at low temperatures.

The magnetic field dependence of the absorption edge discussed in section 3.5 is due to the condensing of the continuum of levels into the first sub-band, so that the shift will be that for $n = 0$, i.e. $\tfrac{1}{2}\hbar\omega_c = \tfrac{1}{2}\hbar Be/m^*$.

It should be noted that as the transitions concerned are from the valence to the conduction band, the correct effective mass to be used in all the expressions is the reduced mass of a hole and electron.

A recent paper by Burstein et al. (1958/9) gives a comprehensive discussion of magneto-optic effects, including the Zeeman effect. Experimental results for Ge include oscillatory magneto-absorption spectra measured with left or right circularly polarized radiation.

CHAPTER 6

EMISSION OF RADIATION FROM SEMI-CONDUCTORS

6.1 Recombination radiation

As discussed in earlier chapters the absorption of photons of energy larger than the forbidden gap of the semi-conductor produces hole electron pairs which give rise to photoconductivity. The quantum efficiency of the process is effectively unity. When at the end of their lifetime the free electrons and holes recombine the reverse process can occur, i.e. the energy given up by the electron on recombining may appear as a photon. As in the absorption process (Chapter 3) the quantum mechanical selection rules must be obeyed. In the majority of semi-conductors the fraction of the recombinations which are radiative will be small, but in InSb for example, such a recombination mechanism may well be the predominant one (Moss, Hawkins and Smith, 1957).

Intrinsic, band-to-band, recombination radiation has now been observed in Ge by Haynes and Briggs (1952a), Newman (1953) and Haynes (1955), in Si by Haynes and Westphal (1956), in InSb by Moss and Hawkins (1956), and in InP, GaSb and GaAs by Braunstein (1955). In addition, longer wavelength radiation has been observed from Ge by Aigrain (1954b). Emission due to avalanche breakdown has been observed by Chynoweth and McKay (1956). Bernard (1958) has reported emission induced by crossed electric and magnetic fields which is effectively the inverse of the P.E.M. effect.

In a semi-conductor which is in equilibrium with its surroundings at room temperature there will be a certain steady rate of photo-excitation determined by the small proportion of the quanta in the room temperature radiation which are sufficiently energetic to give transitions across the forbidden gap. However small this rate is it can be calculated explicitly from standard black body radiation formulae if the absorptive behaviour of the semi-conductor is known.

By the principle of detailed balancing, this rate of excitation in any elementary frequency interval $(d\nu)$ must be exactly equalled by the rate of generation of photons (in the corresponding energy interval) by electron-hole recombination.

The photo-excitation rate is given by

$$R = \int c' q(\nu) K \, d\nu \qquad (6.1)$$

where $q(\nu)$ is the photon density of radiation, c' is the velocity of radiation within the material and K is the absorption constant. Now as $q(\nu)$ increases very rapidly with wavelength the significant part of the integral in equation (6.1) is concentrated around the absorption edge where the absorption index is fairly low and the

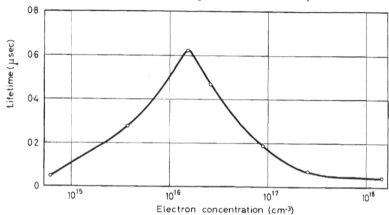

Figure 6.1. Calculated radiation lifetime in InSb. (Landsberg and Moss, 1956; courtesy Physical Society, London)

dispersion small. It is therefore a good approximation to ignore the dispersion and treat the refractive index as constant. We then have

$$c' = c/n \qquad (6.2a)$$

$$q(\nu) = 8\pi\nu^2/(c')^3(\exp h\nu/kT - 1) = 8\pi\nu^2 n^3/c^3(\exp h\nu/kT - 1) \qquad (6.2b)$$

In terms of the Planck function for surface emission,

$$Q(\lambda) = 2\pi c\lambda^{-4}(\exp h\nu/kT - 1)^{-1}$$

which may be taken directly from a radiation slide rule, we have

$$R = 4n^2 \int KQ(\lambda)\,d\lambda \qquad (6.3)$$

This expression has been given by Landsberg and Moss (1956). A more complex relation which may be used if the dispersion becomes important is given by van Roosbroeck and Shockley (1954).

Suppose the equilibrium carrier concentrations are n_0, p_0 and that these are increased slightly in some manner to $n = n_0 + \Delta n$, $p = p_0 + \Delta p$, then the recombination rate becomes

$$R + \Delta R = R(n_0 + \Delta n)(p_0 + \Delta p)/n_0 p_0 \qquad (6.4)$$

92

therefore

$$\Delta R/R = \Delta n/n_0 + \Delta p/p_0 \qquad (6.5)$$

If $\Delta n = \Delta p$

$$\Delta R = R\Delta n(n_0^{-1} + p_0^{-1}) \qquad (6.6)$$

The radiation lifetime τ_R is given by

$$\tau_R = \Delta n/\Delta R = n_0 p_0/R(n_0 + p_0) \qquad (6.7)$$

This expression is a maximum for the intrinsic case, giving

$$\tau_R = n_i/2R \qquad (6.8)$$

The calculated values of the radiation lifetime in InSb for various carrier densities, with allowance for the shift of the absorption edge with impurity concentration, are plotted in *Figure 6.1*. The lifetimes are very short, and represent the maximum lifetimes the carriers can have in this material when all other recombination processes are negligible.

According to Mackintosh and Allen (1955) the effect of degeneracy on the radiative lifetime is negligible.

6.2 SPECTRAL DISTRIBUTION OF RECOMBINATION RADIATION

It has been pointed out by the author (Moss, 1957a) that the spectral distribution of the radiation which is observed experimentally may differ considerably from that generated, depending on the experimental conditions. Suppose the excess carriers are generated photoelectrically by short wavelength radiation which is highly absorbed, so that the carriers can be considered to be generated in a very thin layer on the front surface of a parallel sided thin slab of the material, and that the recombination radiation is observed at the back surface*; the equilibrium rate of photon generation per unit waveband and per unit volume will be

$$r = 4n^2KQ \qquad (6.9)$$

where $R = \displaystyle\int_0^\infty r\, d\lambda$ as in equation (6.3), and K is the photoelectric part of the absorption coefficient. The excess generation rate due to equal increases of Δn in both electron and hole densities will be

$$\Delta r = 2r\Delta n/n_i \qquad (6.10)$$

* For the case of uniform generation throughout the specimen see section 10.7.

93

for an intrinsic specimen. The corresponding total excess generation rate will be

$$\Delta R = 2r\Delta N/n_i \qquad (6.11)$$

where ΔN is the increase in the total number of electrons present.

This radiation will be generated in a solid angle 4π, but due to the usual high value for the refractive index of semi-conductors most of the radiation reaching the back (unilluminated) surface will suffer total internal reflection. For InSb or Ge, for example, where $n = 4$, integration of the Fresnel reflection coefficients over all angles of incidence shows that almost exactly 1 per cent of the radiation is transmitted. By definition all this radiation will be incident on the radiating surface at near-normal incidence and will have travelled through just the thickness t of the specimen and suffered an attenuation $\exp(-K't)$, where K' is now the total absorption coefficient which may differ from K.

The total directly emitted radiation will thus be (for $n = 4$) $10^{-2}\Delta R \exp(-K't)$, of which a fraction γ is assumed to be collected by the optical system. Provided the specimen is such that multiple reflections are unimportant—e.g. the Weierstrass sphere used by Aigrain (1954a)—then the spectral distribution of the emitted radiation will be of the simple form $QK \exp(-K't)$. However, for the usual plate type of specimen, the 99 per cent reflected radiation will travel to the illuminated surface and back to the radiating surface and so on for multiple reflections. Each time one per cent of the remaining intensity will be emitted, provided the surfaces are not perfectly plane parallel. Hence the total collected radiation will be

$$\Delta R_c = 10^{-2}\gamma\{\Delta R \exp(-K't) + (0\cdot99)^2\Delta R \exp(-3K't) \ldots \text{etc.}\} \qquad (6.12)$$

$$= \frac{10^{-2}\gamma\Delta R \exp(-K't)}{1 - (0\cdot99)^2 \exp(-2K't)}$$

$$= 10^{-2}\gamma\Delta R/2 \sinh K't \qquad (6.13)$$

to a good approximation. Hence

$$\Delta R_c/A = KQ/2Q_{\max} \sinh K't \qquad (6.14)$$

where $A = 8 \times 10^{-2}\gamma n^2 Q_{\max}\Delta N/n_i$ is independent of wavelength, $Q_{\max}$ being the peak value of the Planck function Q.

Compared with the simple generation expression of equation (6.9), equation (6.14) has a much greater contribution at long wavelengths and consequently the peak of (6.14) lies at longer wavelengths than the peak of (6.9). Spectral distribution curves for InSb are given in *Figure 6.2*. Curve A is the measured emission, curve B the theoretical generation curve, curve C the theoretical emission curve given by equation (6.14) and curve D the expected emission after correction for the spectral resolution used. The agreement between experiment and theory is seen to be satisfactory.

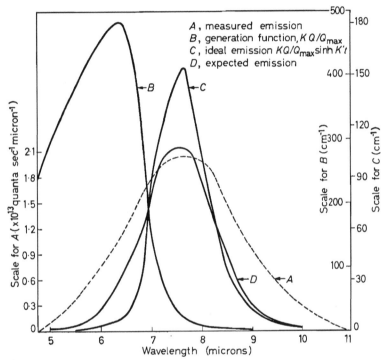

Figure 6.2. *Measured and expected distribution of recombination radiation from* InSb. (*Moss, 1957a; courtesy Physical Society, London*)

The estimation of the percentage of recombinations which are radiative, which equals the ratio of the bulk lifetime to radiation lifetime, τ_B/τ_R, is as follows:

If the incident illumination is such that I photons/sec are absorbed, then

$$\Delta N = I\tau_B \qquad (6.15)$$

Also

$$\tau_R = n_i/2R \qquad (6.16)$$

Substituting the values of ΔN and n_i from these equations in equation (6.14) we obtain

$$\Delta R_c = 10^{-2}\gamma(I\tau_B/\tau_R) \int_0^\infty \frac{QK/Q_{max}\,d\lambda}{2\sinh K't} \bigg/ \int_0^\infty \frac{QK\,d\lambda}{Q_{max}} \quad (6.17)$$

The ratio of the integrals is found from the relative areas under curves B and C, ΔR_c and I are measured and γ is known. Hence τ_B/τ_R is obtained.

Equation (6.17) is modified somewhat if multiple absorption–recombination–absorption processes are considered. Suppose the ratio of the integrals is $1-\beta$, then the fraction β of the radiation must be re-absorbed. Assuming that this absorption produces more electron-hole pairs with unity quantum efficiency then in place of the term $I\tau_B/\tau_R$ we have

$$I\tau_B/\tau_R\{1 + \beta\tau_B/\tau_R + \beta^2\tau_B^2/\tau_R^2 \ldots\}$$

$$= I(\tau_B/\tau_R)(1 - \beta\tau_B/\tau_R)^{-1} \text{ for } \beta\tau_B/\tau_R < 1$$

$$= I\tau_B/(\tau_R - \beta\tau_B)$$

This correction is clearly important when τ_B approaches τ_R.

Using this latter expression in conjunction with equation (6.17) the data of Moss, Smith and Hawkins (1957) give for intrinsic InSb

$$\tau_B/\tau_R = 0{\cdot}21$$

As discussed in the section on InSb, this ratio is in general agreement with measured values of τ_B and calculated values of τ_R.

As all the recombination radiation is not free to leave the specimen, but will be re-absorbed—mainly reproducing hole electron pairs—it is clear that the attainable lifetime in a given specimen can exceed τ_R as defined above. This point has been noted by Dumke (1957a), although this worker exaggerates the situation in stating that there is 'no close connection between τ_R and observable lifetimes'. For the experimental configuration considered above, where excess carriers are produced near to the upper surface, virtually half the radiation—i.e. all that travelling upwards—can reach the surface without absorption, and could therefore escape if the surface was bloomed, for example. With such a configuration therefore, the observed lifetime could not exceed twice τ_R. Also, as in the relevant spectral region the absorption coefficient < 2000 cm^{-1}, and as InSb photoconductive detector elements may well be only ~ 5 μ thick, radiation will reach *both* surfaces fairly easily.

In general, the maximum lifetime attainable under the limitation of radiation recombination will be a complex function of the geometry of the system, the nature of the surfaces, where the excess carriers are generated, diffusion processes and the wavelength dependence of the absorption coefficient.

Both Moss, Hawkins and Smith (1956) and Dumke (1957a) have noted that recombination radiation with multiple absorption generation processes can in theory contribute to the transport of carriers through a semi-conductor. The effect is unlikely to be significant unless the majority of recombinations are radiative and the recombination rate is high—i.e. unless τ_R is very small. Dumke (1957a) shows that the order of magnitude of the diffusion constant (D_{eff}) due to this process is

$$D_{\text{eff}} \sim \langle K^{-2} \rangle / 3\tau_R \qquad (6.18)$$

where $\langle K^{-2} \rangle$ is a weighed function of the absorption, which he suggests could be as large as $\sim 3 \times 10^{-3}$ in cm, sec, units. Hence for InSb, with $\tau_R = 8 \times 10^{-7}$ we have $D_{\text{eff}} \sim 10^3$ cm^2/sec. This figure exceeds the ambipolar diffusivity of intrinsic InSb, so that it is quite possible that this radiation transport process contributes significantly in such material. According to Busch and Schneider (1954) the thermal conductivity of InSb is higher than expected from conventional theories.

6.3 INDIRECT AND DIRECT RECOMBINATION

The theory of sections 6.1 and 6.2 is purely statistical and does not consider the actual details of the recombination processes. By analogy with absorption, both direct and indirect recombinations will occur, the latter involving phonon interaction. As the best known semi-conductors do not have the conduction and valence band extrema at the same points in $\bar{k}$ space (i.e. Si and Ge) the lowest energy transitions will be indirect ones. Indirect transitions in these material may therefore give the more important component of the long wavelength absorption.

If the radiation lifetime is calculated from equation (6.3) using the measured absorption curve, then since the integrand in this equation peaks sharply in the neighbourhood of the absorption edge, it follows that the value of τ_R obtained will approximate to the radiation lifetime for indirect transitions alone, τ_{Ri}.

The following expression for τ_{Ri} has been given by Dumke (1957a):

$$\tau_{Ri} = \frac{2\pi^2 c^2 \tanh \theta/2T}{h^3 E^2 \varepsilon A(n_0 + p_0)} \sum_v m_v^{3/2} \sum_c m_c^{3/2} \qquad (6.19)$$

where E is the energy gap, ε the optical dielectric constant, and the effective mass sums are taken over the band extrema. The factor A is determined from an analysis of the absorption edge. For intrinsic Ge this expression gives $\tau_{Ri} = 2$ secs, as compared with the value of 0·75 sec obtained by van Roosbroeck and Shockley (1954) from the absorption data using essentially equation (6.3).

Dumke (1957a) also calculates the lifetime for direct radiative recombination in Ge, making use of the present knowledge of the band structure of this crystal and the cyclotron resonance data to calculate the optical matrix elements for transitions between the lowest conduction band and both light and heavy hole bands. The result of the calculation is $\tau_{Rd} = 0·3$ secs., so that apparently the direct recombination is more important than the indirect.

The relative importance of radiative recombination and Auger type non-radiative recombination processes are discussed by Bess (1957). This worker shows theoretically that non-radiative recombination, where a hole annihilates a trapped electron (the trap taking up the momentum deficiency and a nearby electron carrying away the excess energy), is the more likely process.

6.4 ELECTROLUMINESCENCE

Since this phenomenon lies in the realm of phosphors rather than semi-conductors, no detailed discussion will be included here. The interested reader is referred to papers by Destriau and Ivey (1955), Kazan and Nicoll (1955) to the proceedings of the 1956 Paris conference on luminescence published in the Journal de Physique et le Radium, Vol. 17, and to recent reviews by Henderson (1958) and Piper and Williams (1958).

6.5 PHOTOELECTRIC EMISSION

The state of theory and experiment in this field has not changed greatly in the last five years, and readers are referred to the review by Moss (1952a).

One new development is that Taft and Apker (1952) have observed that the photoelectric emission in some alkali halides is greatly enhanced for absorption in the exciton bands. Fourie (1954) has observed that the photoemission from a tellurium film, for example, changes on passing a current through the film, whereas no such effect is observed with metal layers. Huntington and Apker (1953) have extended the theory of photoemission to include estimates of the energy dependence of the transition probability.

6.6 EMISSIVITY OF THERMAL RADIATION
FROM SEMI-CONDUCTORS

Although a great deal of work has been carried out on optical absorption in semi-conductors, no corresponding measurements of emission of thermal radiation appear to have been made. As many important semi-conductors have the interesting part of their absorption spectra (i.e. the long wavelength absorption edge) lying quite well into the infra-red region, emission measurements are possible without using excessive specimen temperatures.

There is additional interest in such measurements in order to compare them with observations of recombination radiation, which have now been made on several semi-conductors.

6.6.1 Theory

Consider a unit solid block of material of refractive index n and absorption constant K. Now the number of possible frequencies in a frequency interval $d\nu$ is

$$8\pi\nu^2\,d\nu/v^3$$

where $v = c/n$ is the velocity of propagation of the radiation. The number of photons per unit volume is thus

$$q\,d\nu = 8\pi\nu^2\,d\nu/v^3(\exp h\nu/kT - 1) \qquad (6.20)$$

The absorption in distance dx is $qK\,dx$, and in time dt is $qvK\,dt$. Now this must equal the photon generation rate $r_\nu\,d\nu\,dt$; therefore

$$r_\nu\,d\nu = 8\pi K\nu^2\,d\nu/v^2(\exp h\nu/kT - 1) \qquad (6.21)$$

or

$$r\,d\lambda \equiv -r_\nu\,d\nu = (8\pi Kcn^2\lambda^{-4}\,d\lambda)/(\exp h\nu/kT - 1) \qquad (6.22)$$

where λ is the wavelength *in vacuo*.

This radiation is generated in a solid angle 4π, so that the volume rate of photon generation for unit waveband in a small solid angle $d\omega$ is

$$r\,d\omega = n^2KQ\,d\omega/\pi \qquad (6.23)$$

where

$$Q = 2\pi c\lambda^{-4}/(\exp h\nu/kT - 1) \qquad (6.24)$$

is the tabulated Planck function.

Consider radiation, generated in a horizontal layer of unit area and thickness dx (situated a distance x below the upper surface of a large parallel sided slab of material of thickness X), which is travelling perpendicular to the surface. The radiation generated

99

is $r\,dx\,d\omega$ and that reaching the upper surface is $re^{-Kx}\,dx\,d\omega$. Thus the total primary radiation reaching the upper surface is given by

$$\int_0^X re^{-Kx}\,dx\,d\omega = r[1 - \exp(-KX)]\,d\omega/K \left.\begin{matrix} \\ \\ \end{matrix}\right\} \quad (6.25)$$
$$= n^2Q[1 - \exp(-KX)]\,d\omega/\pi$$

from equation (6.23). Putting $T = e^{-KX}$ for the transmittivity of the slab at the relevant wavelength and R for the reflectivity of the surfaces, then, with allowance for multiple reflections, the radiation generated initially in an upward direction which reaches the upper surface is

$$(r/K)(1 - T)(1 + R^2T^2 + R^4T^4\ldots)\,d\omega.$$

Similarly, the radiation generated downwards which reaches the upper surface is

$$(r/K)(1 - T)(RT + R^3T^3\ldots)\,d\omega$$

Summing these, the total radiation reaching the upper surface is

$$[r(1 - T)\,d\omega]/[K(1 - RT)]$$

and the total radiation emitted from this surface becomes,

$$\frac{r(1 - T)(1 - R)}{K(1 - RT)}\,d\omega = \frac{Q}{\pi}\frac{(1 - R)(1 - T)}{1 - RT}\,d\Omega$$

where $d\Omega = n^2\,d\omega$ is the solid angle into which the radiation is refracted. Now the normal photon emission of a black body is simply $(Q/\pi)\,d\Omega$, so that the emissivity (ε) of the material is given by:

$$\varepsilon = (1 - R)(1 - T)/(1 - RT) \quad (6.26)$$

It should be noted that this expression is independent of refractive index except through R. Also, if the slab is highly absorbing, with $T = 0$, then $\varepsilon = 1 - R$. On the other hand, if $T \to 1$, $\varepsilon \to 0$. The emissivity of a semi-conductor should therefore be fairly high at short wavelengths and should decrease sharply at the absorption edge.

The resemblance of the thermal emission and recombination radiation theories should be noted. In particular, equation (6.25) shows the same spectral dependence as equation (10.5) provided Q is the same—i.e. provided the temperature is the same in both cases. The outstanding experimental difference of course is in the time response, since the emissivity will be governed by thermal time

100

constants whereas recombination radiation can be modulated at frequencies corresponding to typical free carrier lifetimes of $\approx 1\,\mu$ sec.

6.6.2 Results for InSb and Ge

Measurements have been made by the author on indium, antimonide and germanium, (Moss and Hawkins, 1958a). The specimens used were self-supporting and were heated electrically.

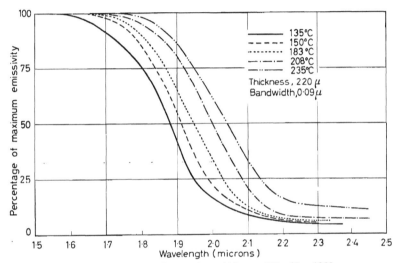

Figure 6.3. Emissivity of germanium. (*Moss and Hawkins*, 1958a; courtesy *Physical Society, London*)

For the germanium work it was necessary to measure at short wavelengths where the Planck function was only $\sim 10^{-5}$ of its maximum value. In this region a liquid air cooled PbS detector was used, in conjunction with a Leiss double monochromator with silica prisms. The results for the relative emissivity are shown in *Figure 6.3*. As expected ε is high at short wavelengths, with a rapid fall at the absorption edge. At the lower temperatures, $\varepsilon \sim 5$ per cent, but at 235°C the value has risen to 0·12 of maximum. According to equation (6.26) this latter figure is equivalent to $K = 4\ \mathrm{cm}^{-1}$.

The emission edge moves to longer wavelengths on heating, the shift of the half maximum point being nearly -5×10^{-4} eV/°C.

The InSb results are given in section 16.9.

101

CHAPTER 7

BORON

7.1 INTRODUCTION

THE difficulty of producing specimens which are either pure or single crystal has greatly handicapped all work on this element.

There is still some doubt about the crystal structure of boron. Laubengayer *et al.* (1943) give lattice constants $a = 8.73$ Å, $c = 10.13$ Å. Lagrenaudie (1953*a*) confirms these values, but considers that half this c value gives the true periodicity of the lattice, with signs of a super-lattice, at $c = 10.13$ Å in some cases. Hoard *et al.* (1951) give a similar result, namely $a = 8.73$, $c = 5.03$, with fifty atoms in the cell.

Various determinations of the thermal activation energy have been made from the slope of log (resistance)/T^{-1} plots. A summary of the results is given in the following table:

Table 7.1—Thermal Activation Energy of Boron

Observer	Nature of Specimens	Activation Energy (eV)
Weintraub (1913)	Fused pieces	1·05–1·25
Warth (1925)	Fused pieces	0·98
Moss (1952*a*)	Films from B_2H_6	0·9, 0·98, 1·0
Uno *et al.*, (1953)	Vacuum reduced	~1·3 (from diagram)*
Lagrenaudie (1953*b*)	Vacuum fusion	Up to 1·2
Lagrenaudie (1953*b*)	Cracking of BBr_3	1·3 (repeatedly)
Bader–Jacobsmeyer (1954)	From B_2H_6	0·86
Greiner–Gutowski (1957)	From BCl_3	1·39 ± 0·05 (repeatedly)

* There is a misprint in the original paper.

In averaging results of this nature preference should be given to the higher values, and it is considered that the best value for the thermal activation energy lies between 1·3 and 1·4 eV.

According to Shaw *et al.* (1953) the holes are slightly more mobile than the electrons.

7.2 TRANSMISSION AND ABSORPTION

Transmission measurements made by Lagrenaudie (1953*b*) on layers made from B_2H_6 are shown in *Figure 7.1a*. For the thicker layers the absorption increases rapidly in the 1·5 μ region. For wavelengths less than 0·7 μ there is no further drop in transmission.

The best absolute determinations of absorption coefficient are

102

those by Morita (1954), using thin films produced by vacuum evaporation. The results are shown in *Figure 7.1b*. An attempt has been made to calculate values from transmission measurements by Lagrenaudie (1953*b*) using rough estimates of thickness by this

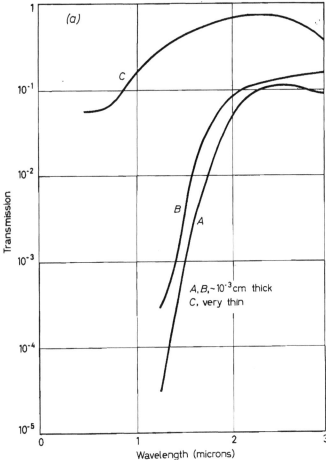

Figure 7.1a. *Transmission of boron layers*

worker. Internal inconsistencies in this paper render the results rather unreliable, but they have been included in *Figure 7.1b*. Also shown is a point calculated from reflection interference fringes measured by Moss (1952*a*). There is no well defined absorption edge on the curve, but there is however a change in

8   103

slope near 1·0 μ—which is shown more distinctly on the curve of extinction coefficient $(\lambda K/4\pi)$ given by Morita (1954). This point may therefore be taken as indicating roughly the position of the absorption edge. Such a wavelength would correspond to an

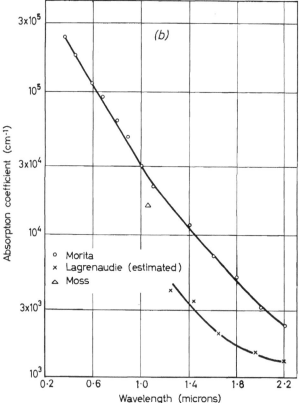

Figure 7.1b. Absorption of boron layers

activation energy consistent with that obtained from electrical measurements.

7.3 REFRACTIVE INDEX

From a value of reflectivity of 0·29 in the region of the absorption edge, Moss (1952a) estimated the index to be $n = 3\cdot3$. The same value of reflectivity has since been obtained by Lagrenaudie (1953a) for visible and near infra-red wavelengths.

At 2·0 μ, Moss (1952a) observed an interference maximum of 54 per cent reflectivity for a boron layer on a glass backing.

Now for a layer of refractive index n between media of indices 1 and μ, equation (1.35) shows the maximum amplitude of reflection to be:—

$$r_{max} = (n^2 - \mu)/(n^2 + \mu) \qquad (7.1)$$

As $r_{max} = (0.54)^{1/2}$ and $\mu = 1.5$, we obtain $n = 3.2$.

Morita (1954) carried out a detailed analysis of the magnitude of reflection and transmission fringes produced by a specimen 0.285μ thick. He found that the index rose rapidly from 1.9 at 0.45μ to a maximum of 3.2 at 1.0μ. This is as would be expected on moving out of the absorption band shown in *Figure 7.1*. Beyond the absorption edge the index fell slowly—again as expected—out to 1.6μ where it was 3.1. Beyond this wavelength however Morita's results show the index increasing again, which is unnatural in view of the low absorption throughout this region.

Lagrenaudie (1953a) measured the dielectric constant ε at 0.5 Mc/s and obtained a value of approximately 13 for a specimen which was cooled to reduce the conductivity. Lagrenaudie suggests that this figure may be rather high and that the true value is perhaps nearer 12. In view of the low accuracy of these determinations, the difference between n^2 and ε is probably not significant.

7.4 Photo-effects

Photoconductivity has been observed in both films and bulk samples of boron. Extensive measurements on boron films made both by vacuum evaporation of powdered boron and by pyrolytic deposition from B_2H_6 are described by Moss (1952a). The photo-current was found to be linear with illuminating intensity up to 0.1 watts/cm^2, and the response time was observed to be negligible in comparison with the 6×10^{-3} sec duration pulses of radiation used in the measurements.

Spectral sensitivity curves for a layer made from B_2H_6 are shown for temperatures of 24°C and 268°C by curves c and d of *Figure 7.2*. It will be observed that the curves are shifted to longer wavelengths on heating, the shift being $dE/dT = -3.5 \times 10^{-4}$ eV/°C. For curve c, $\lambda_{1/2} = 0.97 \mu$.

Curves a and b show sensitivity of crystalline samples measured by Freymann and Stieber (1934) and Lagrenaudie (1953a). It may be noted that the long wavelength fall of sensitivity is much steeper for the bulk samples than for the layers. These latter curves give respectively $\lambda_{1/2} = 0.93 \mu$ and 0.98μ. These three photo-conductive curves thus indicate an average optical activation energy near 1.28 eV.

In view of the value of dE/dT given above, it would be expected that the thermal activation energy (which will be that appropriate to 0°K) would exceed the optical by ~0·1 eV. Within the limits

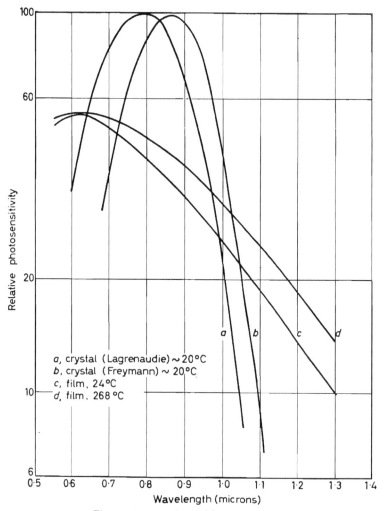

a, crystal (Lagrenaudie) ~ 20°C
b, crystal (Freymann) ~ 20°C
c, film, 24°C
d, film. 268 °C

Figure 7.2. Spectral sensitivity curves for boron

of accuracy of the two energy values such behaviour is seen to occur, and both optical and electrical measurements give a zero temperature energy gap of $1·35 \pm 0·05$ eV.

106

CHAPTER 8

DIAMOND

A REVIEW of the optical and electrical properties of diamond has been given previously by the author (Moss, 1952a). Since then the two most important experimental lines of research on diamonds have been:

(*1*) Study of the counting properties of diamonds. It has been established that there is a strong correlation between the counting efficiency of diamonds and their ultra-violet transparency (Champion and Humphreys, 1957) and also their photoconductivity (Champion and Dale, 1956).

(*2*) Discovery of low resistivity 'semi-conducting' diamonds (Custers, 1954). These diamonds obey Ohm's law, are always *p*-type, and have resistivities ∼100 Ω cm.

On theoretical aspects, Herman (1952) has calculated the energy band structure. He finds that the three lowest conduction bands are degenerate at $\bar{k} = 0, 0, 0$, and that three of the four valence bands are degenerate and have negative curvature at $\bar{k} = 0, 0, 0$. Hence the minimum energy gap—which is found to be close to 6 eV—is probably at this point. More recently the lattice vibration spectrum has been discussed by Raman (1956) and Herman (1957) and the band structure by Brophy (1956) and Dehlinger (1957).

It has been announced recently that diamonds are now being made synthetically in the U.S.A., but so far no measurements on such specimens are available. The problems of synthesis are discussed by Kuss (1956).

8.1 CARRIER MOBILITIES AND SPECTRAL SHIFT

Both the techniques referred to above have been used to obtain better estimates of mobility than were previously available. Allemand and Rossel (1954), from bombardment conductivity measurements have concluded that the mobilities exceed 1000 cm²/V.sec.

Using a semi-conducting diamond Austin and Wolfe (1956) have obtained accurate values of hole mobility from conventional Hall effect measurements. They find $\mu_p = 1550$ cm²/V.sec.

Redfield (1954) has used a novel variation of Hall effect technique which is well suited to the study of high resistivity photoconductors. He finds an electron mobility of 1800 cm²/V.sec with an approximate $T^{-3/2}$ temperature dependence, and a hole mobility $\gtrsim 1200$ cm^2/V.sec.

From these experiments it is concluded that the best values of room temperature mobility are:

$$\mu_e = 1800, \quad \mu_h = 1550 \text{ cm}^2/\text{V.sec.}$$

The ratio of these mobilities agrees well with an earlier measurement by Pearlstein and Sutton (1950).

The elastic constants of diamond have been measured by McSkimin and Bond (1957) using ultra-sonic techniques in the range 20–200 Mc/s. They find

$$c_{11} = 10 \cdot 76 \times 10^{12} \text{ dynes/cm}^2$$

Putting these values in equation (3.19) we obtain

$$C_e = 9 \cdot 4 (m/m_e)^{5/4}, \quad C_h = 10 \cdot 1 (m/m_h)^{5/4}$$

Assuming as a first approximation that $m_e = m_h = m^*$, and $C_e = C_h$, and inserting these values into equations (3.20) and (3.21) we obtain for the calculated broadening and dilatation contributions to the spectral shift for the range 20–130°C, (where $\beta = 1 \cdot 45 \times 10^{-6}$).

$$(\mathrm{d}E/\mathrm{d}T)_b = -0 \cdot 3 (m/m^*)^{3/2} \times 10^{-4} \text{ eV/°C}$$

$$(\mathrm{d}E/\mathrm{d}T)_a = \pm 0 \cdot 3 (m/m^*)^{5/4} \pm 0 \cdot 3 (m/m^*)^{5/4} \times 10^{-4} \text{ eV/°C}$$

The observed shift for temperatures of 20–130°C, (Moss, 1952a), is $\mathrm{d}E/\mathrm{d}T \sim -1 \cdot 6 \times 10^{-4}$ eV/°C so that it is necessary that $m^* < m$. If all the signs are negative, as seems probable from the theoretical band structure (Kimball, 1935) and by comparison with germanium, then we find

$$m_e \doteqdot m_h \doteqdot \tfrac{2}{3} m \tag{8.1}$$

and

$$C_e \doteqdot C_h \doteqdot 18 \text{ eV}$$

giving

$$(\mathrm{d}E/\mathrm{d}T)_a = -0 \cdot 6 \times 10^{-4}, \quad (\mathrm{d}E/\mathrm{d}T)_b = -1 \cdot 1 \times 10^{-4} \text{ eV/°C} \tag{8.2}$$

Some correlation with this shift can be obtained from refractive index data by use of equation (3.25) namely

$$n^4 \propto 1/E$$

Differentiating with respect to temperature we have

$$\mathrm{d}E/\mathrm{d}T = -(4E/n)(\mathrm{d}n/\mathrm{d}T) \tag{8.3}$$

108

For the temperature range 20–130°C interpolation of the results of Ramachandran and Radhakrishnan (1952) gives: $dn/dT = +14 \times 10^{-6}$ average. For a similar temperature range the results of Narasimhan (1955) on the temperature variation of dielectric constant correspond to $dn/dT = 12 \times 10^{-6}$. Hence, using the average of these the calculated total temperature shift from equation (8.3) is,

$$dE/dT = -1 \cdot 2 \times 10^{-4} \text{ eV/°C}$$

This is of the correct sign and approximately the magnitude observed.

Now Ramachandran (1947a) has measured piezo-optic effects in diamond and finds that the refractive index decreases under hydrostatic pressure as follows:

$$\Delta n = -\frac{n^3}{2} \frac{(p_{11} + 2p_{12})}{3} \frac{dV}{V} \tag{8.4}$$

where

$$(p_{11} + 2p_{12})/3 = -0 \cdot 175$$

Hence

$$\Delta n/n = 0 \cdot 5 \, dV/V$$

Putting

$$\frac{(dE)}{(dT)_d} = \frac{dE}{dV/V} \times \frac{dV/V}{dT} = -\frac{4E}{n} \frac{dn}{dV/V} \frac{dV/V}{dT} \tag{8.5}$$

gives the expected dilatation contribution to the shift as

$$(dE/dT)_d = -0 \cdot 5 \times 10^{-4} \text{ eV/°C} \tag{8.6}$$

This is again of the same sign and is approximately equal to the value calculated above (equation 8.2), thus confirming that both negative signs should be taken in equation (3.20).

It is worthy of note that the shift of the absorption edge (Robertson et al. 1934) and the temperature dependence of refractive index become much larger at higher temperatures, both approximately doubling between 20 and 120°C. For the range 20–314°C, the tabulated positions of the absorption edge given by Robertson et al. (1934) are well represented by

$$E(T°C) = 5 \cdot 5[1 - (1 \cdot 9 \times 10^{-5}T) - (1 \cdot 1 \times 10^{-7}T^2)] \text{ eV} \tag{8.7}$$

giving

$$dE/dT = -(1 + 0 \cdot 012T)10^{-4} \text{ eV/°C} \tag{8.8}$$

109

For the refractive index

$$dn/dT = 8 \cdot 5(1 + 0 \cdot 009\,T°\text{C})10^{-6} \qquad (8.9)$$

so that

$$-(4E/n)dn/dT = -0 \cdot 8(1 + 0 \cdot 009\,T°\text{C})10^{-4}\,\text{eV}/°\text{C}$$

There is thus reasonable agreement between the absolute magnitude of the measured shift and its value calculated on the basis of $n^4E = \text{constant}$, and agreement between the temperature dependencies which is probably as good as the accuracy of the measurements.

In terms of equations (3.19), (3.20) and (3.21) much of this temperature dependence of the shift may arise in the following way. Wedepohl (1957) has found that the hole mobility

$$\mu_p \propto T^{-2 \cdot 8}$$

in contrast to the usual $T^{-1 \cdot 5}$ law, so that equation (3.19) gives

$$C_h{}^2/\rho u^2 = T^{1 \cdot 3}$$

Hence from equation (3.21)

$$(dE/dT)_b \propto T^{1 \cdot 3}$$

Over the temperature range 0–300°C this can be replaced quite accurately by

$$(dE/dT)_b \propto 1 + 0 \cdot 006\,T°\text{C}$$

Further, assuming that ρu^2 is constant, we have

$$C_h{}^2 \propto T^{1 \cdot 3} \quad \text{or} \quad C_h \propto 1 + 0 \cdot 003\,T°\text{C}$$

Now from the results of Krishnan (1946) the expansion coefficient is markedly temperature dependent, and can be represented by

$$dV/V\,dT = 2 \cdot 6(1 + 0 \cdot 01\,T°\text{C})$$

so that from equation (3.20), assuming. that both C_e and C_h have the same dependence,

$$(dE/dT)_d \propto 1 + 0 \cdot 013\,T°\text{C}$$

Hence as this latter term is about half the broadening term, the expected temperature dependence of the total shift is

$$dE/dT \propto 1 + 0 \cdot 009\,T°\text{C}$$

which is close to the observed behaviour.

110

8.2 ACTIVATION ENERGIES

Clark *et al.* (1956) have measured the optical absorption of many diamonds. Defining the absorption edge as the point where a change of slope of the absorption spectrum is first observed in type II diamonds, these workers conclude that the activation energy is $5 \cdot 4 \pm 0 \cdot 03$ eV at 80°K. Champion and Humphreys (1957) also report spectra for a large number of diamonds. They find that for several specimens the wavelength at which observable absorption first appears is as low as 2264 Å, or 5·5 eV. These energy values are very close to earlier estimates from optical, electrical and photoelectric work (Moss, 1952a). Custers and Neal (1957) emphasize that the u.v. transmission limit depends to some extent on the diamond thickness. A recent analysis of the absorption edge by Clark (1958) in terms of indirect transitions, gives a room-temperature energy gap of 5·4–5·5 eV, with phonon energies of 0·09 eV and ∼0·23 eV. The theoretical figure for the cohesive energy of diamond is given as 6 eV by Schmidt (1953).

In the infra-red spectrum of type IIb diamonds Clark *et al.* (1956) find a very strong absorption line at 3·6 μ. Austin and Wolfe (1956) find the same prominent line in their semi-conducting diamond—its peak absorption coefficient being 8 cm⁻¹ at 3·57 μ. which corresponds to an impurity activation energy of 0·35 eV.

From the temperature dependence of the resistivity and Hall constant of their semi-conducting diamond, Austin and Wolfe (1956) obtain an impurity activation energy of 0·38 eV. Almost the same activation energy was found for an 80 Ωcm specimen by Dyer and Wedepohl (1956). These latter workers found for four specimens that the room temperature conductivity was related directly to the strength of the infra-red absorption bands. Energy values near 0·35 eV are deduced from measurements by Custers (1955), Brophy (1955) and Leivo and Smoluchowski (1955).

Wedepohl (1957) has carried out a thorough experimental investigation of semi-conducting diamonds and has made a detailed analysis of the results. Three of his five specimens gave identical values for activation energy, namely 0·34 eV, and this is probably the most reliable value yet available. On calculating the effective mass from the carrier concentration data, these three specimens gave the same value, namely $m_h{}^* = 0 \cdot 25 \, m$. This is smaller than the figure estimated from the shift data above (equation 8.1).

The above values of impurity activation energy are close to those for a hydrogen-like model of a singly ionized impurity in

111

a medium of dielectric constant equal to that of diamond. Equation (3.28) gives 0.42 eV for $m_h = m$ and 0.28 eV for $m_h = 2/3m$. The extent of this experimental and theoretical agreement indicates that the acceptor impurity in the semi-conducting diamonds is probably a group III element—aluminium is likely (Brophy, 1956).

The main i.r. absorption band at $4.8\ \mu$, which is present in all types of diamond, has been shown to be independent of impurities (up to $10^{18}\ \mathrm{cm^{-3}}$) and disorders (up to $10^{19}\ \mathrm{cm^{-3}}$) by Collins and Fan (1954). They conclude therefore that it is a lattice absorption band and show that its temperature variation is consistent with this conclusion.

Further work on the i.r. absorption of diamond has been published by Stephen (1958b), and Raal (1958). For very long wavelengths the transmission of type II diamond is so good that it has been used for windows in Golay cell detectors. Semi-conducting diamonds however show free carrier absorption in the 10–24 μ region, particularly at elevated temperatures where the acceptors are fully ionized. With allowance for the fact that $h\nu \sim kT$, these measurements give the effective mass of holes as $m_h \sim \frac{1}{2}m$.

In the visible region, Dyer and Matthews (1958) have shown that the absorption and fluorescence emission spectra are very similar.

8.3 CARBON–SILICON MIXTURES

Although arbitrary mixtures of carbon and silicon have not yet been reported, some measurements are available for the specific compound SiC.

For alpha SiC Choyke and Patrick (1957a) found an energy gap of 2.86 eV at $300°$K by transmission measurements, with a temperature dependence of -3.3×10^{-4} eV/$°$C. These workers concluded that the transitions were indirect, involving emission or absorption of a phonon of 0.09 eV energy. For pure cubic SiC, Lely and Kröger (1956/58) find the absorption edge ~ 2.8 eV.

From resistance–temperature measurements Racette (1957) obtains an intrinsic activation energy of 3.1 ± 0.2 eV, in good accord with the optical values.

The dielectric constant at low temperatures is 10.2 (Hofman et al., 1957). This is considerably greater than the square of the refractive index which is ~ 6.7, indicating a moderate degree of ionicity. From analysis of the Restrahlen reflectivity, Lely and Kröger (1956/8) deduce that the effective ionic charge is $e^* = 0.94e$.

CHAPTER 9

SILICON

9.1 GENERAL PROPERTIES

A GREAT deal of research work has been carried out on silicon in the past few years, partly because (with germanium) it is a relatively simple monotomic semi-conductor on which to make fundamental measurements, and partly stimulated by its applications for radar detector and mixer crystals, transistors and solar batteries.

It crystallizes in the diamond lattice with $5 \cdot 2 \times 10^{22}$ atoms/cm³,

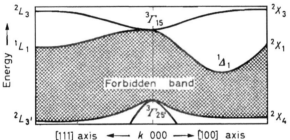

Figure 9.1. Energy bands in silicon. (*Herman*, 1955b; courtesy *Taylor and Francis, London*)

the nearest neighbour separation being $2 \cdot 35$ Å and lattice constant $5 \cdot 431$ Å. The coefficient of linear expansion at room temperature is $\beta = 4 \cdot 15 \times 10^{-6}$. The technique of growing high purity single crystals is well established (Pfann, 1957; Wilson, 1958), the limit of purity at the present day corresponding to ~ 1000 Ω cm resistivity, although occasional specimens $> 10^4$ Ω cm have been reported (Cronemeyer, 1957).

Recent review articles on the optical and photoelectric properties of pure and doped silicon have been given by Burstein, Picus and Sclar (1956), Fan, Shepherd and Spitzer (1954/6) and Fan (1956).

9.2 BAND STRUCTURE

The band structure of silicon has been investigated theoretically by Yakama and Sugita (1953), Jenkins (1954) and by Herman (1954, 1955a,b) whose results are shown in *Figure 9.1*. The top of the valence band lies at $\bar{k} = 0$, and the energy surfaces there are approximately spherical. The band is degenerate, and the different curvature of the two energy surfaces correspond to two types of

113

holes with different mobilities. For the slow holes cyclotron resonance data gives a mass of $0.46m$–$0.56m$ depending on the crystal direction, and for the fast holes $0.16m$. The weighted average for use in conductivity formulae, assuming both bands to have the same relaxation time, is given by

$$\bar{m} = (m_1^{3/2} + m_2^{3/2})/(m_1^{1/2} + m_2^{1/2}) \tag{9.1}$$

so that

$$\bar{m} \sim 0.39m$$

(Dexter et al. 1954a). There is also a third surface which is slightly separated (~ 0.03 eV) due to spin-orbit splitting (Dunlap, 1957).

The lowest point in the conduction band does not occur at $k = 0$, but somewhere along the [100] crystal direction. The energy surfaces are six equivalent ellipsoids, for which the effective mass along the major axis (i.e. [100] direction) is $m_L = 0.97m$, and the transverse mass, $m_T = 0.19m$. This surface is not degenerate, and the average effective mass for use in conductivity equations is given by

$$3/\bar{m} = 1/m_L + 2/m_T \tag{9.2}$$

i.e. $\bar{m} \sim 0.26m$ (Dresselhaus et al., 1955a).

Values close to these average values for holes and electrons have been estimated by Spitzer and Fan (1957b) from the long wavelength reflectivity of p and n type silicon.

The minimum separation between conduction bands—which is the thermal activation energy—is 1.1 eV, but as seen from *Figure 9.1* the minimum *vertical* transition is ~ 2.5 eV. Direct transitions will thus only be possible with visible photons, any infra-red absorption must arise from indirect transitions.

9.3 ELECTRICAL PROPERTIES

Cronemeyer (1957) has measured the mobilities of the carriers in Si, using very high purity material. His values, and drift mobility values determined by Ludwig and Watters (1956) for 300°K are given in *Table 9.1*. Neither mobility obeys the simple $T^{-3/2}$ temperature dependence. The law is approximately $T^{-2.5}$ for both carriers.

Table 9.1

	Electrons	Holes
Hall mobility (C)	1560 cm²/V.sec	345 cm²/V.sec
Drift mobility (C)	1360	510
Drift mobility (L and W)	1350	480

Morin and Maita (1954*b*) have analysed detailed Hall and conductivity data taken over the temperature range 10–1100°K, and have found the energy gap to be 1·21 eV at 0°K. Also from purely electrical data they estimate that the temperature dependence of the activation energy is about $-3\cdot6 \times 10^{-4}$ eV/°C, so that their

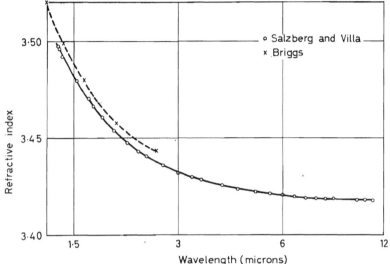

Figure 9.2. Refractive index of silicon

estimated room temperature activation energy is 1·10 eV. A more accurate estimation of this temperature dependence term, using the cyclotron resonance values of effective mass in the conduction formulae, is $-4\cdot0 \times 10^{-4}$ (Burstein and Egli, 1955), giving a room temperature activation energy of 1·09 eV.

From the absolute value of the conductivity and mobility, Brooks (1955) gives an energy gap of 1·09 eV.

9.4 Refractive index and dielectric constant

Very accurate measurements of refractive index of single crystal silicon have been made by Salzberg and Villa (1957) in the region of good transmission, using a large prism (3 × 3 cm faces). Their results are shown in *Figure 9.2*, with some earlier prism measurements by Briggs (1950) for comparison. The latter worker was able to measure down to 1·05 μ, where the index was 3·565. At 11 μ, where dispersion seems to have reached negligible proportions, the index is $n_0 = 3\cdot4176$, indicating a dielectric constant $n_0{}^2 = 11\cdot68$.

The dielectric constant at radio frequencies has been measured

115

directly by Dunlap and Watters (1953) by doping with approximately 10^{-5} per cent of gold and then cooling with liquid nitrogen to produce material of high resistivity ($\sim 10^{10}$ Ω cm). For frequencies of 500 c/s–30Mc/s the value was $\varepsilon = 11\cdot7$, which is in excellent agreement with the value from the refractive index.

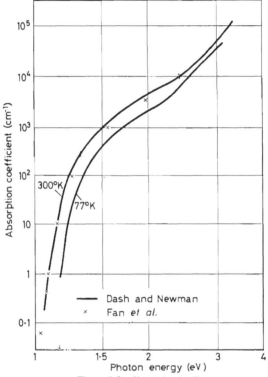

Figure 9.3. Absorption in silicon

Dash (1955) has observed birefringence in the infra-red, due to strains introduced during crystal growth. The temperature dependence of the refractive index and dielectric constant has been discussed by Antoncik (1956). Cardona *et al.* (1958) find $dn/n\ dT = 7 \times 10^{-5}$ per °C, with negligible pressure dependence.

9.5 INTRINSIC ABSORPTION IN SILICON

9.5.1 Short wavelength absorption

Detailed transmission measurements on pure single crystal silicon have been reported by Fan, Shepherd and Spitzer (1954/6), showing absorption coefficients ranging from $< 0\cdot1$ to $\sim 10^3$ cm^{-1}.

116

By preparing extremely thin specimens, Dash and Newman (1955) have succeeded in extending such measurements to absorption levels of 10^5 cm^{-1}. These results are shown in *Figure 9.3*.

In conjunction with *Figure 9.1* these results are interpreted as showing the onset of indirect transitions at approximately 1·1 eV at room temperature. At high absorption levels there is an increase of absorption, indicating the threshold of direct transitions, at about 2·5 eV. This is shown rather more clearly on the low temperature curve of *Figure 9.3*.

Macfarlane and Roberts (1955b) have made transmission measurements at temperatures down to 20°K and have analysed the absorption very near the edge in terms of non-vertical transitions. They find that their data can be represented quite well by equation (3.11), with an energy gap of 1·08 eV at 300°K, and a phonon energy of 0·052 eV. From their deduction of the size of the phonons involved they estimate that the conduction band minima occur at a momentum value which is about two-thirds of that of the zone edge. From a recent analysis—including exciton effects—these workers obtain 1·123 eV gap, phonons of 212,670, 1050 and 1420°K and an exciton binding energy of 0·007 eV.

Values for the optical energy gap at room temperature by various workers are summarized in *Table 9.2*. For low temperatures Zwerdling* has quoted the precise value of 1·2053 eV.

Table 9.2—Activation Energy of Silicon at Room Temperature

Observer	eV
Macfarlane and Roberts	1·08
Dash and Newman	1·06
Fan, Shepherd and Spitzer	1·05
Bell Telephone Laboratories	1·090*
R.R.E. Malvern	1·123* at 291°K

* Recent unpublished work. (Physical Society Meeting, Malvern, and Rochester Conference, 1958)

A comparable value for the activation energy can be obtained from electrical data. As stated in section 9.3, such data give the zero-temperature energy gap $E_0 = 1·21$ eV. If the gap shift with temperature is assumed to be linear, with $dE/dT = -4 \times 10^{-4}$ eV/°C, then $E_{290} = 1·09$ eV. However, if as suggested by Macfarlane and Roberts (1955b) the gap varies as (temperature)², as for germanium and diamond, i.e. $E = E_0 - \alpha T^2$ where

$$dE/dT = -2\alpha T = -4 \times 10^{-4} \text{ eV/°C}$$

then
$$E_{290} = 1·15 \text{ eV}$$

117

These values are then in general accord with the figures of *Table 9.2.*

In spite of the high values of absorption coefficient reached by Dash and Newman (1955), the total absorption shown there is still quite trivial when compared with the total amount required to explain the refractive index of silicon. The total absorption shown in *Figure 9.3* is

$$\int_{0\cdot37}^{\infty} K \, d\lambda \doteqdot 1$$

whereas the relation

$$n_0 - 1 = \frac{1}{2\pi^2} \int_0^{\infty} K \, d\lambda$$

(section 2.43) shows that

$$\int_0^{\infty} K \, d\lambda = 48$$

Measurements over the wavelength range $0\cdot2$–1 μ, by Pfestorf (1926) by analysis of the reflection of polarized light, show that there is a pronounced peak in $2nk$ at $0\cdot35$ μ, the maximum value of $2nk$ being 25. It has been shown by the author (1952a) by use of equation (2·26a) that integration of Pfestorf's $2nk$ values gives the correct value for the real part of the dielectric constant.

9.5.2 *Shift of absorption edge*

Direct measurement of the temperature dependence of the photon energy for a given absorption level by both Dash and Newman (1955) and Fan, Shepherd and Spitzer (1956) give $-4\cdot5 \times 10^{-4}$ eV/°C. With allowance for a slight change in absorption magnitude with temperature Fan, Shepherd and Spitzer (1956) conclude that the true temperature dependence of the edge is given by $dE/dT = -4\cdot0 \times 10^{-4}$ eV/°C. Almost the same value has also been deduced from purely electrical measurements (see section 9.3), and from the temperature dependence of the spectral sensitivity of a p–n junction photocell (Bardeen and Shockley, 1950). Macfarlane and Roberts (1955b) find that the shift is smaller at low temperatures, perhaps for the same reasons as suggested for diamond (see section 8.1).

The pressure dependence of the absorption edge has also been studied, and it seems to be established now that the wavelength of the edge *increases* with pressure. Warschauer, Paul and Brooks (1955) give $dE/dp = -2 \times 10^{-12}$ eV/dyne cm^{-2}, while a slightly later publication by Paul and Pearson (1955), working in the

118

range of intrinsic conductivity at 250°C, gives $dE/dp = -1·5 \times 10^{-12}$ eV/dyne cm^{-2}. The value obtained by Paul and Pearson (1955) from the pressure dependence of the intrinsic conductivity (at the same temperature) is also $-1·5 \times 10^{-12}$ eV/dyne cm^{-2}. From section 9.1 the volume expansion coefficient is $12·4 \times 10^{-6}$ per °C, and the compressibility (Fan, Shepherd and Spitzer, 1956) is $\chi = -0·98 \times 10^{-12}$ per dyne cm^{-2}. Hence the energy shift due to thermal dilatation of the lattice is

$$(dE/dT)_d = (3\beta/\chi)\ dE/dp \doteqdot 0·25 \times 10^{-4}\ \text{eV/°C}$$

From the drift mobility values given in section 9.3 and equation (3.19) we can evaluate the interaction constants for electrons and holes. They are given by:

$$C_e{}^2 = 20(m/m_e)^{5/2}, \qquad C_h{}^2 = 56(m/m_h)^{5/2}$$

For electrons the appropriate mass average to be used is (Brooks, 1955),

$$m_e^{-5/2} = \tfrac{1}{3}(m_L m_T{}^2)^{-1/2}(m_L^{-1} + 2m_T^{-1}) \qquad (9.3)$$

giving $C_e = 20$ eV. For holes, the average mass depends on the relative proportions of carriers in the 'light' and 'heavy' hole bands. Its value is probably near that quoted in section 9.2 for use in conductivity equations, i.e.

$$m_h \doteqdot 0·39m$$

Hence

$$C_h \doteqdot 24\ \text{eV}$$

Thus from equation (3.20)

$$(dE/dT)_d = 8·3 \times 10^{-6}(\pm 24 \pm 20)$$

In order to obtain a small positive shift, it is clear that we must use $+C_h$ and $-C_e$, giving

$$(dE/dT)_d \doteqdot 0·3 \times 10^{-4}\ \text{eV/°C}$$

This value agrees reasonably well with the experimental value, and thus lends support to this conclusion that the conduction and valence band extrema move in opposite directions on compression.

Using the above values of C_e and C_h we can calculate the contribution to the shift by broadening, from equation (3.21), obtaining

$$(dE/dT)_b = -2·4 \times 10^{-4}\ \text{eV/°C},$$

if we use the average effective masses given in section 9.2.

The total shift expected theoretically is therefore

$$(dE/dT)_a + (dE/dT)_b \doteqdot -2 \times 10^{-4}$$

This is of the correct sign and correct order of magnitude, but only half the observed shift. However, it is pointed out by Fan (1956) that the masses to be used in equation (3.21) for $(dE/dT)_b$ are characteristic of the whole energy bands, whereas the effective masses which are measured by the conventional methods are determined only by the band edges, and that the values to be used in equation (3.21) should therefore be much higher than the average effective masses. He suggests that masses roughly equal to free electron masses are likely. With such masses and the values of C_e and C_h given above, the calculated total shift would indeed be near to the experimental value.

9.5.3 Free carrier absorption

Free carrier absorption in silicon has been studied for p-type material by Briggs (1950) and Fan and Becker (1951), whose results show that the absorption is roughly proportional to carrier concentration and to (wavelength)^2. Measurements have been extended to 35 μ by Spitzer (see Fan, 1956) who finds that the absorption coefficient increases smoothly as λ^2 over the whole range beyond 10 μ. There is no evidence of an absorption band due to transitions between the different valence bands in the wavelength region expected from the estimated band separation of 0·035 eV.

Kahn (1955) has analysed this data and finds good agreement with equation (2.30b), provided a suitable value of effective mass is used in this equation. For most of the specimens for which he carried out the analysis, the resulting effective mass was between 0·3m and 0·36m. These values are in reasonable accord with the average mass given in section 9.2.

Measurements on n-type material have been made by Spitzer and Fan (1957a), using specimens with up to 10^{19} cm^{-3} free electrons at room temperature. The absorption is proportional to λ^2 from ~5 μ to the limits of the measurements, which was 45 μ in some cases. For a specimen with 6×10^{18} cm^{-3} free electrons, the absorption coefficient rose to 2800 cm^{-1} at 20 μ. Analysis of the results shows that quantitative agreement with theory is obtained if an average value of effective mass of $m^* = 0·3m$ is used.

Spitzer and Fan (1957a) have also observed a weak absorption band superimposed on the λ^2 spectrum. Subtraction of the λ^2

term gives the results shown in *Figure 9.4.* The band peaks near
2·3 μ, and for a specimen with $5 \cdot 5 \times 10^{17}$ cm⁻³ antimony atoms the
maximum absorption was only 1·5 cm⁻¹, but it increased in direct
proportion to the free carrier concentration. This absorption may
be due to indirect transitions to higher conduction levels, possibly
to the minimum at $\tilde{k}000$ shown in *Figure 9.1.*

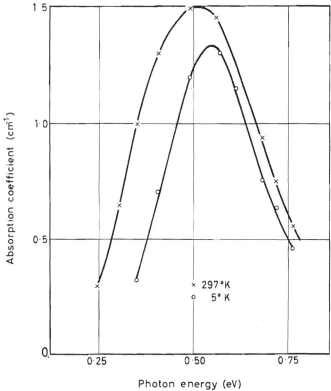

*Figure 9.4. Free electron absorption band in silicon. (Spitzer and Fan,
1957a; courtesy American Physical Society)*

9.5.4 *Lattice absorption bands*

Absorption bands have been observed in silicon by Collins and
Fan (1954) and Burstein, Picus and Sclar (1956) which are quite
independent of both impurity concentrations and lattice imperfections, and these bands are therefore ascribed to the silicon lattice.

The absorption in the region up to 25 μ is shown in *Figure 9.5.*

121

The two main bands are at 8·9 μ and 16·5 μ, and have peak absorption coefficients of 4 cm⁻¹ and 10 cm⁻¹ respectively, at room temperature. The temperature dependence of the absorption is the same as for the calculated mean square displacement of the atoms, and is therefore assumed to arise from the vibrating atoms. These silicon absorption bands appear to correspond to the bands in type II diamonds.

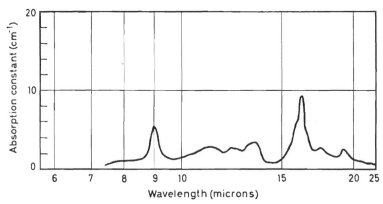

Figure 9.5. Lattice absorption in silicon. (*Lax and Burstein*, 1955a; courtesy *American Physical Society, New York*)

9.6 INTRINSIC PHOTOCONDUCTIVITY

The main technical interest in intrinsic photoconductivity in Si lies in its use in *p–n* junction form for the direct conversion of sunlight to electrical power in the 'solar battery'. As discussed in section 4.6.4 the optimum energy gap for such a device is ∼1·3 eV, where the theoretical efficiency reaches 25 per cent.

The energy gap of silicon is not much less than this optimum and so far it has produced the best performance figures. At elevated temperatures the optimum activation energy becomes slightly greater (Halstead, 1957), so that in such circumstances silicon may not be quite so good.

Design of solar batteries has been discussed by Cummerlow (1954), Chapin *et al.* (1954), Prince (1955) and Maslakovets *et al.* (1956). Units with efficiencies exceeding 10 per cent have already been made, in contrast to GaAs where the best efficiency so far is 6·5 per cent (Jenny *et al.*, 1956).

Recent developments in silicon photo-voltaic devices (Prince and Wolf, 1958) give solar conversion efficiencies of 14%.

9.7 RECOMBINATION RADIATION

Recombination radiation from silicon has been observed by Haynes and Briggs (1952b). The spectral distribution curve is symmetrical, with a peak at 1·12 μ and falling to 50 per cent on the long wavelength side at 1·21 μ. At low temperatures, Haynes and Westphal (1956) and Haynes (1958) have found additional structure sensitive radiation which is attributed to recombination of excess carriers with un-ionized donors or acceptors. The latter worker evaluates the transverse and longitudinal optical phonons as 1390 and 970°K, and corresponding acoustical phonons as 190 and 640°K.

Using the theory of section 6.1 and the absorption data of *Figure 9.3*, the calculated radiation lifetime for silicon is $\tau_R \sim 4$ secs. This value is some 10^4 times greater than the experimental values for good pulled crystals (Bittman and Bemski, 1957), so that direct electron-hole recombination is not the predominant recombination mechanism. Calculations of the relative probabilities of various mechanisms, by Sclar and Burstein (1955), show that such behaviour is theoretically reasonable.

9.8 ABSORPTION IN DOPED SILICON

Considerable work has been carried out on the optical absorption and photoconductivity of silicon doped with group III and group V impurities, and with gold. This work is greatly stimulated by the technical interest in long wavelength infra-red detectors using photo-effects in doped Si or Ge.

The activation energies which have been determined optically are shown in *Table 9.3*, the donor energies being relative to the bottom of the conduction band and the acceptor energies relative to the top of the valence band. Most of these values are from Collins and Carlson (1957), from a review of shallow impurity levels in Si and Ge by Kohn (1957), and from Hrostowski and Kaiser (1958).

Table 9.3—Impurity Ionization Energies in Silicon (eV)

Acceptors		Donors	
B	0·046	P	0·045
Al	0·067	As	0·053
Ga	0·071	Sb	0·043
In	0·16	Bi	0·071
Au	∼0·5	Mn	0·53
Zn	0·31	Au	∼0·7
Cu	0·49	Fe	0·55 and 0·75
		Cu	∼0·9 (0·24 from valence band)

123

There has been some uncertainty about the energy levels in gold doped silicon. The activation energies for gold in the above table are the $\lambda_{1/2}$ values from photoconductive spectral response curves on p and n type material by Collins, Carlson and Gallagher (1957). The donor level agrees with the estimate from electrical data by Taft and Horn (1954) of a donor level 0·33 eV above the valence band. For Tl, Shulman (1957) quotes a thermal activation energy of 0·26 eV, but no optical data are available.

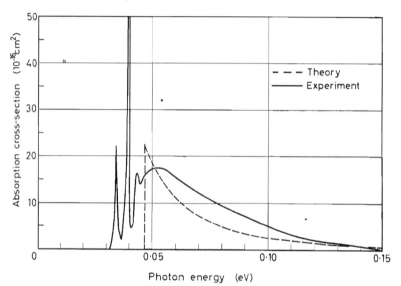

Figure 9.6. Absorption in boron doped silicon. (Burstein, Picus, Henvis and Lax, 1955; courtesy American Physical Society, New York)

Detailed studies of the absorption of B, Al, Ga and In impurities have been made by Burstein, Picus and Sclar (1956). At very low temperatures all three materials show sharp lines corresponding to excitation levels of the bound holes. The results for boron are shown in *Figure 9.6.* At slightly higher energies the ionization continuum—corresponding to the onset of photoconductivity—begins.

It was thought originally that the positions of the lines in *Figure 9.6* could be fitted by a simple hydrogenic spectrum in a medium of dielectric constant ε, namely

$$E = 13 \cdot 56 \ (m^*/m\varepsilon^2)(1 - 1/n^2) \qquad (9.4)$$

where $n = 2$, 3, 4, etc. However, Kohn (1957) concludes that this agreement was purely fortuitous.

More extensive structure has been measured by Hrostowski and Kaiser (1958). The energy levels of the lines found by these workers are given in *Table 9.4*.

Table 9.4—*Energies of Lines in Acceptor Spectra at 4°K*

	meV							
Boron	30·16	34·52	38·40	39·65	41·46	42·13	42·74	43·85
Aluminium	54·91	58·57	64·08	64·99	66·90	67·55	68·51	
Gallium	58·26	61·80	67·15	68·26	70·68	71·39	72·28	
Indium	142·06	145·73	149·75	150·88	153·36	153·96	155·41	

The last entry in each row is estimated to be very near to the optical ionization energy for that impurity. Kohn (1957) has carried out detailed calculations of the energy levels expected, and obtained good agreement with experiment for the excited states. His results for the ground state are not so good, possibly because the smaller orbits lie too near the donor ion for the use of a simple Coulombic potential in a dielectric to be valid.

The absolute value of the measured absorption cross-section of boron centres at $h\nu = 0.046$ eV is 15×10^{-16} cm². The calculated absorption cross-section for photo-ionization of a hydrogenic system at frequencies near the ionization limit is given by Burstein, Picus and Sclar (1956) as

$$\sigma = \frac{8\cdot3 \times 10^{-17}}{nE_i}(m/m^*)(E_i/h\nu)^{8/3} \qquad (9.5)$$

where n is the refractive index. For boron where $E_i = 0.046$ eV, and $m^*/m = 0.45$ as before, this gives $\sigma = 12 \times 10^{-16}$ cm², in good agreement with the experimental value.

At liquid helium temperatures the widths of the absorption lines (in *Figure 9.6*) are $\sim 10^{-3}$ eV. It has been shown by Sampson and Margenau (1956) that at negligible temperatures a total broadening of $1\cdot6 \times 10^{-3}$ eV would be expected theoretically, arising mainly from inter-action of the bound hole with lattice vibrations. Lax and Burstein (1955b) show that the increase in line width between helium and nitrogen temperature may also be explained on the basis of the interaction of the lattice vibrations with the trapped carrier. Broadening of the lines also occurs if the carrier concentration is increased above 10^{16} cm⁻³, to such an extent that for concentrations $\sim 10^{18}$ cm⁻³ (at 20°K) the line structure disappears (Newman, 1956).

The ionization energies of P and As donor levels in *Table 9.3* are in close agreement with theoretical values determined by Kohn (1955, 1957).

Absorption by oxygen in silicon has been studied by Kaiser *et al.* (1956) and by Hrostowski and Kaiser (1957). The latter workers find a band at 1106 cm⁻¹ at room temperature, which becomes a triplet with peaks at 1135·5, 1127·9 and 1121 cm⁻¹ on cooling to 50°K or below. There is also a band at 515 cm⁻¹. The effect of heat treatment on the optical properties of Si has been investigated by Kaiser (1957).

9.9 PHOTOCONDUCTIVITY IN DOPED SILICON

As the activation energies of the impurities in silicon are in general very small, it is necessary to cool to low temperatures to prevent

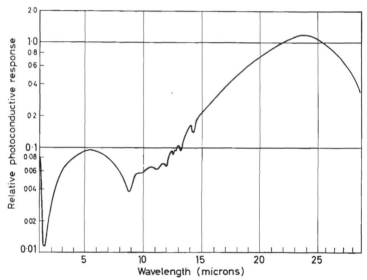

Figure 9.7. Spectral response of silicon containing boron and indium. (Burstein, Picus, Henvis and Lax, 1955; courtesy American Physical Society, New York)

thermal excitation of the centres before photo-effects become important. For good photosensitivity it seems necessary to cool until $kT < \Delta E/20$, and much of the experimental work so far done has been at liquid helium temperatures.

A spectral sensitivity curve for silicon containing both indium and boron is shown in *Figure 9.7*. The short wavelength band of sensitivity due to the indium centres is seen to fall to half value at

$\lambda_{1/2} = 8 \cdot 2$ μ, or $0 \cdot 15$ eV. The long wavelength band due to the boron centres extends to 28 μ ($0 \cdot 045$ eV) for $\lambda_{1/2}$.

Of the doping elements listed in *Table 9.3*, Zn, In and Au have activation energies sufficiently high to be operated at liquid air temperatures. For Si doped with Zn, Carlson (1957) finds $\lambda_{1/2} = 3 \cdot 5$ μ. Indium doped silicon shows particular promise for a relatively long wavelength infra-red detector with this convenient amount of cooling. Newman (1955) and Blakemore (1956) have studied photoconductivity in this material at temperatures up to 77 or 90°K.

The observed absorption cross-section at the ionization limit for In doped Si is only $0 \cdot 7 \times 10^{-16}$ cm^{-2}, so that even with impurity concentrations as high as 10^{17} cm^{-3}, thicknesses of several mm are required to absorb all the incident radiation. Use of such thick specimens is not of course conducive to high photosensitivity.

For copper doped silicon, transitions to the donor level near the valence band give $\lambda_{1/2} = 4 \cdot 5$ μ at 77°K, (Collins and Carlson, 1957).

9.9.1 Design of doped photoconductive detectors

In the fundamental design of doped photoconductive detectors, there is (in principle, at least) a free choice of the following parameters: (1) doping concentration, (2) donor or acceptor impurities, (3) operating temperature and (4) activation energy.

The density of free electrons in a specimen containing N_d donors and N_a acceptors, where $N_d > N_a$ is given by

$$n(n + N_a)/(N_d - N_a - n) = (N_c/g) \exp\left(-\Delta E_d/kT\right) \qquad (9.6)$$

where ΔE_d is the donor activation energy, g the degeneracy of the level (usually taken to be 2, (Fan, 1955)) and N_c is the effective density of states in the conduction band. A similar equation may be written for holes if $N_a > N_d$.

Now for maximum photosensitivity it is necessary to keep this concentration of carriers in the unilluminated specimen as small as possible (see Chapter 4). In general it is required to make the activation energy small and keep the operating temperatures high, so that the exponential term is determined by these parameters, both of which tend to make it large. Also it is necessary to keep the concentration of the operative centre (the donor in this case) fairly high in order to keep the absorption constant relatively large.

However, the value of n in equation (9.6) can be greatly reduced by adjustment of N_a—i.e. by counter doping with acceptors. Consider a suitable example where $N_d = 10^{17}$ cm^{-3}, $N_c/g = 10^{19}$ cm^{-3},

$\Delta E_d = 0{\cdot}1$ eV, and $kT = 0{\cdot}008$ eV (90°K); then

for (1) $\qquad N_a = 0, \quad n^2 = (N_c N_d/g) \exp(-\Delta E/kT)$ $\qquad$ (9.7)

or $\qquad\qquad\qquad n = 2 \times 10^{14}$ cm^{-3}

for (2) $\qquad\qquad\qquad N_a \doteqdot N_d$

$$n = \frac{(N_d - N_a)}{g N_a} N_c \exp(-E/kT) \qquad (9.8)$$

$$= 10^{13} \text{ cm}^{-1} \text{ for } N_a = 0{\cdot}8 \, N_d$$

Thus fairly extensive counter doping can reduce the 'dark' carrier concentration greatly. It should be noted that, as the fraction of centres thermally excited is normally negligible, $N_d - N_a$ gives the density of occupied donors, so that the absorption coefficient is proportional to this difference, rather than to N_d. Hence, for a given absorption coefficient, there is no advantage in *very* accurate compensation, it is simply necessary to make N_a fairly large in equation (9.8).

Some choice is also available in the density of states parameter, since it will not be the same for the conduction band and the valence bands. We have

$$N_c = 2(2\pi k T/h^2)^{3/2} \delta (m_L m_T^2)^{1/2}$$

where the number of band minima is $\delta = 6$. Hence

$$N_c = 1{\cdot}13 \times 2(2\pi m k T/h^2)^{3/2}$$

and

$$N_v = 2(2\pi k T/h^2)^{3/2}(m_1^{3/2} + m_2^{3/2})$$

$$= 0{\cdot}46 \times 2(2\pi m k T/h^2)^{3/2}$$

Thus for a given impurity concentration equation (9.8) shows that an acceptor impurity will give only 40 per cent of the 'dark' carrier concentration that a donor impurity will give. For silicon therefore, acceptor doping will give the better photosensitivity, provided impurities of the desired activation energy are available.

Recent developments in this field have been discussed by Burstein and Picus (1958). Davis (1958) has shown that silicon doped with impurities which have deep lying levels makes a good crystal counter for α particles.

CHAPTER 10

GERMANIUM

10.1 GENERAL PROPERTIES

SINCE the last war, germanium has been the subject of more intensive and more extensive research work than any other semiconductor, or for that matter any other conductor. It fully justifies the statement that 'the best known conductor is now only a semiconductor'. The impetus for this work has stemmed mainly from its great technological value as *the* transistor material, and from the fact that it is the most reasonable of the monatomic semiconductors for fundamental work.

At an early stage in the work, the technique of growing large single crystals of high purity was mastered, and it is on such good quality material that virtually all recent research work has been carried out. These techniques have been reviewed by Pfann (1957) and Cressell and Powell (1957).

Ge crystallizes in the diamond lattice with $4 \cdot 5 \times 10^{22}$ atoms/cm³. The lattice constant at 20°C is $a = 5 \cdot 6575$ Å, and the expansion coefficient is $5 \cdot 9 – 6 \cdot 6 \times 10^{-6}$ per °C (Straumanis and Aka, 1952).

The intrinsic resistivity at 27°C is 47 Ω cm, corresponding to a carrier density of $2 \cdot 6 \times 10^{13}$ cm⁻³ each of holes and electrons. The best material now available has $\sim 10^{12}$ cm⁻³ impurities, so that it can be considered as truly intrinsic at room temperature and a little below.

Recent reviews on various aspects of germanium have been published by Herman (1955a), Fan (1955) and Dunlap (1957).

10.2 BAND STRUCTURE

The results of the theoretical investigation of the band structure of germanium by Herman (1954, 1955a, b) are shown in *Figure 10.1*.

The maximum of the valence band lies at the centre of the zone, and the surfaces there are approximately spherical. The situation is complicated however by the fact that there are three distinct bands to be considered. Two of these are degenerate levels which meet at the $\bar{k}000$ point, but differing in curvature and hence in effective mass. The mass of the 'heavy' holes is $\sim 0 \cdot 32m$ (for an accurate value the significant degree of anisotropy must be considered, Dexter, Zeiger and Lax, 1954b) and that of the 'light' holes $0 \cdot 042m$. The third band is separated by spin-orbit splitting.

From infra-red absorption work the separation is 0·28 eV, which is sufficiently large for any carriers in this band to be ignored at ordinary temperatures. The energy surfaces of this band are truly spherical, with a mass of 0·077m. The slow holes are mainly responsible for the conduction processes, but the small percentage of light holes have such high mobility ($\sim$12,000 cm²/V.sec) that their effect is important in transport processes in the presence of a magnetic field (Willardson *et al.*, 1954). The lowest of the

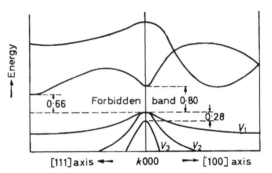

Figure 10.1. Energy bands in germanium. (Herman, 1955b; courtesy Taylor and Francis, London)

various conduction band minima shown in *Figure 10.1* are those along the [111] directions. The minima are probably at the zone faces, so that there are four equivalent minima, (Macfarlane *et al.*, 1957). The energy surfaces there are very elongated ellipsoids, the longitudinal and transverse masses being (Kohn, 1957):

$$m_L = 1 \cdot 6m, \qquad m_T = 0 \cdot 081m$$

Some information on the band structure of both Ge and Si has been obtained from soft x-ray absorption and emission spectra by Tomboulian and Bedo (1956).

10.3 ELECTRICAL PROPERTIES

The drift mobilities of electrons and holes in germanium are (Dunlap, 1957):

$$\mu_e = 9 \cdot 1 \times 10^8 T^{-2 \cdot 3}, \quad \mu_h = 3 \cdot 5 \times 10^7 T^{-1 \cdot 67} \text{cm}^2/\text{V.sec}$$

or at room temperature,

$$\mu_e = 3800, \qquad \mu_h = 1800$$

Due to the 1 or 2 per cent of 'fast' holes, higher values are obtained for μ_h from measurements of Hall effect (Morin, 1954), or Faraday effect at *cm* wavelengths, (Rau and Caspari, 1955).

From a detailed study of Hall effect and conductivity over a wide temperature range, the energy gap at $0°K$ was found by Morin and Maita (1954*a*) to be 0·785 eV. Analysis of this data by Burstein, Picus and Sclar (1956), with use of the effective masses given in section 10.2, shows that the energy gap is temperature dependent, the variation being -4×10^{-4} eV/°C. Thus the room temperature energy gap is 0·66 eV.

10.4 REFRACTIVE INDEX AND DIELECTRIC CONSTANT

Very accurate measurements of the refractive index of germanium in the transparent infra-red region have been made by Salzberg and

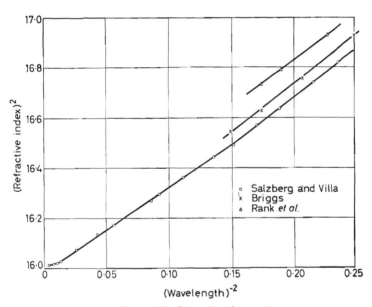

Figure 10.2. Dispersion of germanium

Villa (1957), 8. These workers used a prism with faces $4 \times 4·5$ cm² and 11·8° refracting angle cut from a single crystal. Their results for 27°C are shown in *Figure 10.2*, together with some earlier prism measurements by Briggs (1950).

Also shown in *Figure 10.2* are some results by Rank *et al.* (1954). These measurements were made interferometrically, and although carried out with high precision they are suspect because of the

131

interpretation of the results. It is shown in section 1.5 that the presence of a linear term in the dispersion of a material cannot be detected by measuring the wavelength separation of adjacent fringes (or by counting fringes between given wavelengths as done by Rank *et al.*, 1954).

If the true refractive index is given by $n = n' + \beta\lambda$ where n' includes all terms except the linear one, then equation (1.44) shows that the parameter obtained by counting fringes is n' not n. As by comparison with InSb (section 16.8) β will be expected to be negative, the interferometric value will be too high. This is in agreement with the relative values shown in *Figure 10.2*. Furthermore, the interferometric value should be too large by an amount proportional to λ. The differences between the data of Salzberg and Villa (1957) and Rank *et al.* (1954) at 2·1 μ and 2·4 μ show exact proportionality to λ. It thus seems probable that the dispersion includes a small linear term and that consequently the results given by Rank *et al.* (1954) are slightly in error.

It will be seen from *Figure 10.2* that at the longest wavelength (16 μ), the dispersion has become almost negligible. Extrapolating the straight line to $\lambda = \infty$ gives $\varepsilon = 15\cdot98$, $n = 3\cdot994$. Dispersion at long wavelengths due to free carriers has been observed by Spitzer and Fan (1957*b*). With 4×10^{18} cm^{-3} free electrons, the index fell to 2·0 at 24 μ.

The dielectric constant of germanium has been measured at radio frequencies by Dunlap and Watters (1953). These workers doped the Ge with $\sim10^{-5}$ per cent of gold and cooled to 77°K in order to produce very high resistivity material. At 1 Mc/s they obtained the value $15\cdot8 \pm 0\cdot2$, which, within the limits of accuracy of the experiment, coincides with the square of the infra-red refractive index.

The dispersion of the dielectric constant at radio frequencies was first observed by Benedict (1953). This dispersion is due to free carriers and is given by equation (2.30*a*). In the subsidiary relation for the lattice collision frequency (equation 2.31) namely

$$g = e/\mu m^*$$

Benedict used the geometric mean of the Hall and drift mobilities. He found the effective mass of holes to be: $m_h^* = (0\cdot3 \pm 0\cdot13)m$. More recent work by Goldey and Brown (1955) gives

$$m_h^* = (0\cdot30 \pm 0\cdot05)\,m, \quad m_e^* = (0\cdot09 \pm 0\cdot04)m,$$

where some of the uncertainty arises from different ways of averaging effective masses.

The temperature dependence of the refractive index has been measured by Rank *et al.* (1954) who find that it increases with temperature, the change being $dn/dT = 5.25 \times 10^{-4}$ at 2.25μ.

On the basis of the relation between energy gap and refractive index given in equation (3.25) namely,

$$En^4 = \text{constant} \qquad (10.1)$$

we expect

$$dE/E\,dT = -4\,dn/n\,dT = -5.1 \times 10^{-4}$$

or

$$dE/dT = -3.4 \times 10^{-4}\,\text{eV}/°\text{C}$$

This is of the correct sign and is in quite good quantitative agreement with the measured temperature dependence of the activation energy quoted in sections 10.3 and 10.5.2.

Both temperature and pressure dependence of the dielectric constant and refractive index have been discussed by Cardona, Paul and Brooks (1958) who find $dn/n\,dp = -5 \times 10^{-7}$ per Kg cm^{-2} and $dn/n\,dT = 7 \times 10^{-5}$.

10.5 Intrinsic absorption in germanium

10.5.1 Short wavelength absorption

Direct transmission measurements of the absorption coefficient of germanium have been made by Dash and Newman (1955) using slices of single crystals. By preparing specimens less than 1μ thick they were able to measure K up to 10^5 cm^{-1}. Their results for the longer wavelength region are shown in *Figure 10.3*. In addition to the absorption starting at quantum values equal to the thermal activation energy, the curves show clearly that at $K \sim 100$ the absorption increases sharply due to the onset of direct transitions. From their data, Dash and Newman (1955) deduce that the thresholds for the onset of direct and indirect transitions are as given in *Table 10.1*.

Table 10.1. *Threshold Energies for Transitions in* Ge

	300°K	77°K
Indirect	0.62 eV	0.72 eV
Direct	0.81 eV	0.88 eV

A very precise value for the energy gap for direct transitions has been deduced by Zwerdling and Lax (1957) from studies of the oscillatory behaviour of the absorption near this photon energy in the presence of large magnetic field. They obtained a room temperature gap of 0.803 ± 0.001 eV; (this value is used in *Figure 10.1*). They were also able to evaluate the effective mass in the conduction band minimum at $\tilde{k}000$, the value being $0.042m$.

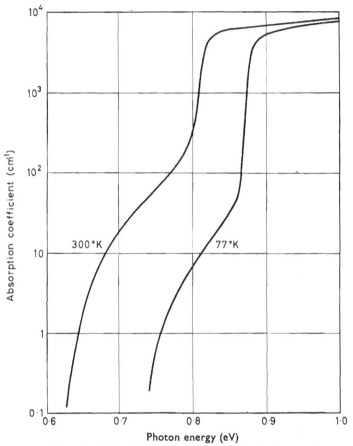

Figure 10.3. Absorption in germanium. (Dash and Newman, 1955; courtesy American Physical Society)

Using a field of 40,000 G and a temperature of 4°K, Zwerdling, Lax and Roth (1957,8) have observed fine structure in the oscillatory magneto-absorption spectrum. They obtain direct and indirect gaps of 0.898 and 0.744 eV.

The absorption edge at low K levels has been analysed by Macfarlane and Roberts (1955a) by fitting their measured values to equation (3.11). They find good agreement, using a phonon of 0·022 eV, and obtain an energy gap at 300°K of 0·65 eV. A more recent treatment by Macfarlane *et al.* (1957) shows that the low level absorption near the edge in Ge is best analysed in terms of indirect transitions involving *two* phonons. These are a 320°K longitudinal phonon and a 90°K shear phonon. The energy dependence associated with the 320°K phonon is interpreted in terms of formation of excitons of 0·005 eV energy. At 291°K the average value for the optical energy gap is found to be 0·670 eV.

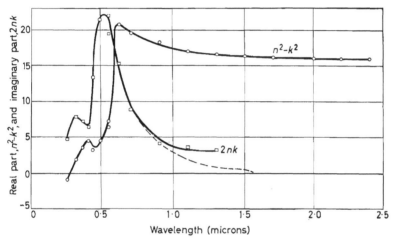

Figure 10.4. Real and imaginary parts of the dielectric constant of germanium.
(Avery and Clegg, 1953; courtesy Physical Society, London)

The optical constants in the short wavelength, highly absorbing region, have been measured by Avery and Clegg (1953). These workers analysed the reflection of polarized radiation from a natural crystal face found on a germanium single crystal. Their results, plotted in *Figure 10.4*, show that the peak of $2nk$ lies at 0·5 μ, and as expected from section 2.4 the maximum in the real part coincides approximately with maximum slope of the imaginary part. The total measured absorption of *Figure 10.4* has been integrated according to equation (2.26a), to find the zero frequency refractive index. The value obtained is $n_0{}^2 = 11·3$, which is only 70 per cent of the known value of 16. Hence either the values of $2nk$ in *Figure 10.4* are rather too low, or there is considerable additional absorption beyond the limit of the measurements—i.e., below 0·25 μ. The optical constants have also been measured by

135

Archer (1958). For the waveband used (0·3–0·7 μ) the integrated value of $2nk$ is only half that required to give $\varepsilon_0 = 16$, but by reasonable extrapolation of the data equation (2.26) can be satisfied.

Koc (1957) has measured the reflectivity at short wavelengths, finding a main peak near 0·6 μ and a smaller peak around 0·4 μ, which is in harmony with the data of *Figure 10.4*.

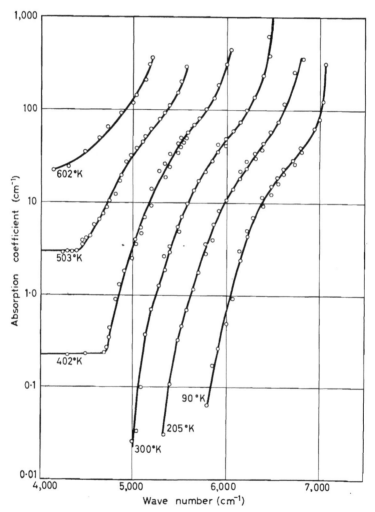

Figure 10.5. Absorption in germanium. (Fan, Shepherd and Spitzer, 1954/56; courtesy John Wiley, New York)

136

10.5.2 Shift of absorption edge

A series of transmission measurements at different temperatures by Fan, Shepherd and Spitzer (1956) are shown in *Figure 10.5*. The absorption edge moves to shorter wavelengths on heating, as for silicon. If the curves of *Figure 10.5* are displaced both to the left and upwards, they are found to superimpose over a wide range of absorption values. Using this procedure the shift from *Figure 10.5* is found to be $dE/dT = -4.4 \times 10^{-4}$ eV/°C. The curves of Dash and Newman (1955) in *Figure 10.3*, if similarly displaced to superimpose at low absorption levels, give $dE/dT = -4.5 \times 10^{-4}$ eV/°C. As pointed out by Tomura and Otsuka (1955) these values are similar to the temperature dependence of the energy gap deduced from electrical measurements, in section 9.3. It may be noted that the shift of the threshold for direct transitions in *Figure 10.3*, between 77°K and 300°K is equivalent to a linear shift of -3×10^{-4} eV/°C. However, very accurate determinations of the position of the edge for direct transitions have been made by Zwerdling, Lax and Roth (1957, 8) from high resolution grating spectra of the oscillatory magneto-absorption at various temperatures. They find

298°K, 0·803 eV;   77°K, 0·890 eV;   4°K, 0·897 eV

These results show that the activation energy is a purely quadratic function of temperature, namely

$$E = 0.897 - 1.06 \times 10^{-6} T^2$$

so that at room temperature,

$$dE/dT = -6.3 \times 10^{-4} \text{ eV/°C}$$

Macfarlane and Roberts (1955a) and Macfarlane *et al.* (1957) show that the position of the absorption edge varies quadratically with temperature below 200°K. Such non-linear movement of the edge has also been found for diamond, and as suggested in Chapter 8, it may be associated with a non-linear thermal expansion co-efficient for the crystal. Above 200°K, Macfarlane *et al.* (1957) find that the shift is constant, its value being $dE/dT = -3.7 \times 10^{-4}$ eV/°C. Such behaviour has been predicted by Antoncik (1955).

The pressure shift of the absorption edge has been measured by Fan, Shepherd and Spitzer (1956) and Warschauer *et al.* (1955), both groups giving values of $(8 \pm 1) \times 10^{-12}$ eV/dyne cm^{-2}. From the ratio of compressibility to thermal expansion coefficient, this shift is equivalent to a thermal dilatation term:

$$(dE/dT)_d = -1.5 \times 10^{-4} \text{ eV/°C}$$

137

This is in order of magnitude agreement with the theory of section 3.3 when the interaction constants calculated from equation (3.19) are used with equation (3.18). However, because of the highly anisotropic electron mass, the presence of light and heavy holes and the influence of differing scattering mechanisms, a detailed comparison of theory and experiment is very difficult. The matter has received considerable attention from Brooks (1955).

10.5.3 Free carrier absorption

Measurements of free carrier absorption in germanium for wavelengths up to 12 μ have been made by Fan and Becker (1951). After subtracting the residual (frequency independent) absorption the remainder was found to be approximately proportional to (wavelength)^2 over the range 5 μ–12 μ.

These data have been analysed by Kahn (1955) on the basis of equation (2.30b). For n type materials he finds that most of the data are consistent with an effective mass:

$$m^* = 0.12m - 0.16m$$

This compares favourably with the average value obtained from cyclotron resonance work (section 9.2) namely

$$3/m_{av} = 1/m_L + 2/m_T \qquad (10.2)$$

or

$$m_{av} = 0.12m$$

Measurements on n type materials have been extended to 38 μ by Fan, Spitzer and Collins (1956). Again the absorption coefficient increases approximately as (wavelength)^2 over a wide range of wavelengths and temperatures. However, slight deviations from direct proportionality are observed, and these workers give a detailed theoretical analysis of the data. They point out that the use of a constant damping factor (g in equation 2.30b) for all the electrons cannot, in general, be a good approximation. It should be satisfactory for photon energies $<kT$, and in this region therefore equation (2.30b) should apply, but at higher photon energies their detailed treatment of the problem shows that the frequency dependence of the absorption should be given by the following two terms:

(a) Lattice scattering term: absorption $\propto \lambda^{3/2}$

(b) Impurity scattering term: absorption $\propto \lambda^{7/2}$

Quantitative agreement between theory and experiment is obtained using an effective mass of approximately $0.1m$. In measurements up to 30 μ on impure type Ge, Spitzer and Fan (1957b) find that although the dielectric constant decreases as λ^2, as expected from equation (2.30a), the absorption coefficient increases as $\lambda^{2\cdot9}$. From this dispersion data they calculate an effective mass of $0\cdot15m$.

Absorption in p-type Ge is dominated by interband transitions

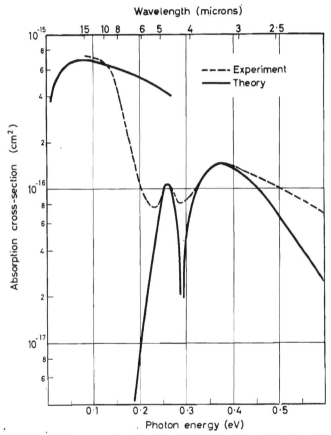

Figure 10.6. Free hole absorption in germanium. (Burstein, Picus and Sclar, 1956; courtesy John Wiley, New York)

arising from the three distinct sub-bands shown in *Figure 10.1*. Measurements by Briggs and Fletcher (1953)—who used carriers injected at a p–n junction—and by Kaiser *et al.* (1953), indicate that there are three absorption bands. As shown in *Figure 10.6*, the peaks of the bands are at 3·4 μ, 4·7 μ and ∼15 μ. These wavelengths are assumed to correspond to the following transitions in *Figure 10.1*,

(a) $3\cdot4$ μ (0·37 eV), band V_3 to V_1

(b) $4\cdot7$ μ (0·27 eV), band V_3 to V_2

(c) 15 μ ($\sim$0·08 eV), band V_2 to V_1

Burstein, Picus and Sclar (1956) point out that the matrix element for transitions between the bands vanishes at $\tilde{k} = 0$, and that the transition probability should increase initially as $\tilde{k}^2$. If all transitions have their maxima at about the same $\tilde{k}$ value, then the sum of the photon energies for transitions (b) and (c) will equal that for (a). This is seen to be approximately the case. The detailed calculations by these workers of the theoretical absorption spectrum are shown by *Figure 10.6* to compare well with the experimental results. By extrapolating the experimental results for the individual bands to zero or infinite wavelength (as appropriate to the theoretical expressions for absorption cross section) they show that at $\tilde{k} = 0$, bands V_1 and V_2 coincide and the separation between bands V_1 and V_3 is 0·28 eV.

Kahn (1955) has also carried out a detailed treatment of interband transitions in Ge, and reaches essentially the same conclusions as given above. He points out that the main factor responsible for the decrease in absorption at high $\tilde{k}$ values is the low density of free holes. As this density will be exponentially dependent on temperature, then the widths of the absorption bands should show strong temperature dependence. Such behaviour is observed experimentally. Newman and Tyler (1957) have measured the absorption over a very wide range of acceptor concentrations. No saturation of the absorption takes place, K increasing approximately linearly with concentration up to 8000 cm^{-1} (at $h\nu = 0\cdot5$ eV) for $1\cdot5 \times 10^{20}$ cm^{-3} acceptors. With decreasing concentration the 0·27 eV band sharpens and becomes relatively more pronounced.

The absorption by injected carriers was first discussed by Lehovec (1952) and Gibson (1953). The technique has been extended by Harrick (1956) to study the distribution of excess carriers in Ge specimens, and by Arthur *et al.* (1956) and Gibson and Granville (1957) to make practical infra-red or microwave modulators. Equation (2·30b) shows that modulation can be achieved by varying either N, the number of carriers (i.e. by injection), or g, through decreasing the mobility by use of very high electric fields.

Small changes of reflectivity (up to $\frac{1}{2}$ per cent) on injection have been observed by Filinski (1957) and treated theoretically by Sosnowski (1957).

10.5.4 Lattice absorption bands

The absorption due to lattice vibrations has been measured for pure Ge by Lax and Burstein (1955a), whose results are shown in *Figure 10.7*. The main band is at 29 μ where the room temperature

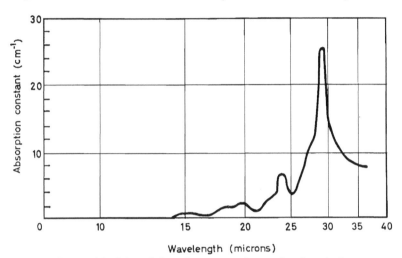

Wavelength (microns)

Figure 10.7. Infra-red absorption spectrum of germanium due to lattice vibrations. (Lax and Burstein, 1955; courtesy American Physical Society, New York)

absorption coefficient reaches 26 cm⁻¹. Collins and Fan (1954) have shown that these bands are insensitive to impurities up to 10^{18} cm⁻³ and to imperfections up to 10^{19} cm⁻³, and have also shown that the temperature dependence of the absorption is in accordance with the calculated mean square thermal displacement of the atoms, thus confirming that the absorption is due to the vibrating lattice.

10.6 INTRINSIC PHOTOCONDUCTIVITY

Early work on photo-effects on germanium has been covered previously by the author (1952a). In particular, it was shown that, for wavelengths from 1 μ to the absorption edge, the quantum efficiency was unity—i.e. one hole-electron pair was produced per absorbed photon. Such measurements have been extended into the ultra-violet region by Koc (1957), who has found that the quantum efficiency is substantially unity down to 0·5 μ. For

141

shorter wavelengths however, it increases linearly with photon energy, the constant of proportionality being 2·5 eV per hole/electron pair. The same constant of proportionality was found for the sensitivity to x-rays by Drahokoupil *et al.* (1957).

Tauc (1954) determined the optical activation energy by the

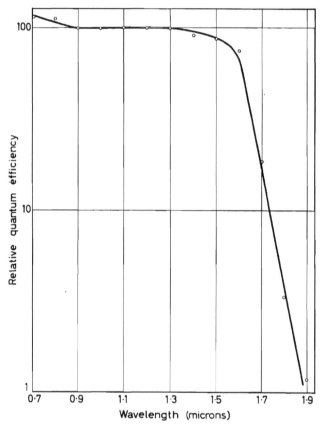

Figure 10.8. P.E.M. spectral dependence in germanium. (*Moss*, 1953*c*; courtesy *Physical Society, London*)

method of 'photoelectric lines' (see Moss, 1952*a*) and obtained the value 0·62 ± 0·02 eV at 300°K.

Spectral sensitivity measurements using the photoelectromagnetic effect for near-intrinsic material are shown in *Figure 10.8.* It will be seen that $\lambda_{1/2} = 1·64$ μ. From the theory of the

P.E.M. effect (equation 4.83) this wavelength corresponds to the condition, $KL = 1$. From *Figure 10.3*, $K = 70$ cm^{-1} at $1 \cdot 64$ μ, so that the ambipolar diffusion length is $L = 0 \cdot 14$ mm. A detailed analysis of photo-conductive and P.E.M. spectral measurements at temperatures of 20–300°K is given by Macfarlane and Roberts (1956/8).

Other P.E.M. measurements have been reported recently by Kikoin and Bykovskii (1956) and Buck and McKim (1957). The latter workers studied the effect over a wide range of surface recombination velocities (values of st/D from $0 \cdot 2$ to >10, in the notation of Chapter 4) and illuminating intensities.

Carrier lifetimes up to $2 \cdot 5$ m sec have been determined from photoconductivity measurements by Stevenson and Keyes (1954). In more recent work lifetimes as long as 6 m sec (Cressell and Powell, 1957) and surface recombination velocities as low as 10 cm/sec have been observed. These lifetimes are still well below the calculated radiation lifetime, which, using equations (6.3) and (6.8), and the data of *Figure 10.3*, is $\tau_R = 300$ m sec.

10.7 RECOMBINATION RADIATION

Early studies of the radiation from Ge due to recombination of excess carriers were made by Haynes and Briggs (1952a) and Newman (1953). The emission measured was a single band with peak intensity at $1 \cdot 8$ μ.

More recent measurements by Haynes (1955), using very thin specimens, have shown that a second peak exists near $1 \cdot 5$ μ. These two peaks correspond to indirect and direct transitions respectively.

The rate of generation of radiation per unit waveband and per unit volume is given by equations (6.9) and (6.11) as:

$$\Delta r = 2r \Delta n / n_i = 8n^2 Q K \Delta n / n_i \qquad (10.3)$$

where $\Delta n / n_i$ is the relative concentration of excess carriers and

$$Q = 2\pi c \lambda^{-4} \exp\left(-h\nu/kT\right) \text{ as } h\nu \gg kT \qquad (10.4)$$

The specimens used by Haynes (1955) were much thinner than a diffusion length, so that the excess carriers may be considered to be uniformly distributed. Hence the total radiation reaching the surface will be:

$$\int_0^t \Delta r \exp\left(-Kx\right) \mathrm{d}x = 8n^2 Q[1 - \exp\left(-Kt\right)]\Delta n / n_i \qquad (10.5)$$

This expression shows that the emission should increase linearly

143

with Δn. Such behaviour was found by Newman (1953). It also shows that if K is large enough for $Kt \gg 1$, then the emission is independent of K—i.e. it will have no structure characteristic of the material. This is the reason why earlier measurements did not show increased emission corresponding to the onset of direct transitions. From *Figure 10.3*, it is only for $K \sim 10^3$ cm^{-1} that direct transitions are important, so that specimens of thickness

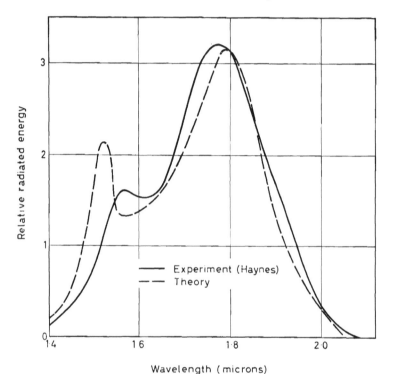

Wavelength (microns)

Figure 10.9. Recombination radiation from germanium

$t \sim 10$ μ are needed. The sample used by Haynes (1955) was of about this thickness.

Using equation (10.5) and the data of *Figure 10.3*, with $t = 15$ μ, the computed emission spectra of *Figure 10.9* is obtained. The general agreement with the measurements of Haynes (1955) is seen to be satisfactory. At $1\cdot5$ μ the value of Q in equation (10.5) is only $\sim 10^{-9}$ of its peak value and is thus a very steep function of temperature. Thus slight differences in ambient temperature for

144

the measurements of *Figures 10.3* and *10.9* could well explain the amount of discrepancy which is observed.

Newman (1957) has observed an additional emission peak at 0·5 eV when deformed material is used. A similar band was reported by Benoit ·à la Guillaume (1956) who ascribed it to transitions following trapping of carriers.

Measurements of the emission of thermal radiation of Ge are described in section 6.6.

10.8 DOPED GERMANIUM

Considerable work has now been carried out on the properties of germanium doped with many different elements. The activation energies of group III and group V donors are all near 0·01 eV, but there are significant differences existing between them. As this energy corresponds to an absorption edge or photoconductive threshold at $\sim$120 μ, little optical or photoelectric work has been carried out on these particular doping agents. Nearly all the other elements investigated show higher activation energies, and in consequence their infra-red properties have been investigated more widely. With the exception of one of the three levels for gold, all the elements with multiple levels give acceptors. Present knowledge on impurity activation energies in Ge is summarized in *Table 10.2*.

Table 10.2. Energy Levels of Impurities in Germanium*

Element	eV from conduction band	eV from valence band	Element	eV from conduction band	eV from valence band
Li	$\sim$0·01 D		Cu	0·26	0·04, 0·32
P	0·0120 D		Cd		0·05, 0·16
As	0·0127 D		Pt	0·20	0·04
Sb	0·0096 D		Au	0·20, 0·04	0·05 D, 0·15
B		0·0104	Ni	0·31	0·22
Al		0·0102	Co	0·30	0·25
Ga		0·0108	Fe	0·27	0·34
In		0·0112	Mn	0·37	$\sim$0·1
Zn		0·030, 0·10	Te	0·11D, 0·3D	
Ag	0·09, 0·28	0·13	Se	0·14D, 0·3D	

* These data are mainly from Dunlap (1957). Kohn (1957) and Tyler (1958)

10.8.1 Absorption in doped germanium

The absorption at low temperatures of germanium doped with various elements has been studied by Burstein, Picus and Sclar (1956) and Fan (1956) and his co-workers. For group III donors Fan *et al.* (1958) find the main lines to be at 111, 99, 151 and 113 μ for P, As, Sb and Bi respectively.

For both Zn and In doping, the absorption curves are similar to the free hole absorption spectra shown in *Figure 10.6*. At helium

temperatures the curves show fairly sharp absorption peaks at
3·5 μ and 3·8 μ respectively. The In peak is thus at ∼0·02 eV
lower photon energy than Zn, in accord with the data of *Table
10.2*. The measurements do not extend to sufficiently long wavelengths to show the optical ionization energy, the absorption
(in the case of In doping) still increasing at 30 μ.

Copper doped germanium has been studied by Kaiser and Fan

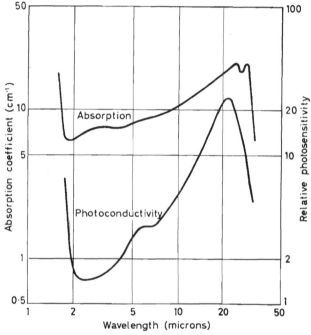

Figure 10.10. Absorption and photoconductivity in copper doped germanium.
(Fan, 1956; courtesy Physical Society, London)

(1954) whose results at helium temperatures are shown in *Figure
10.10*. Here the absorption reaches a maximum within the range
of measurements (at ∼22 μ). At the longest wavelengths K falls
rapidly, indicating an absorption edge near 30 μ, which is in good
agreement with the activation energy determined electrically. The
magnitude of the absorption near the long wavelength cut-off
is in reasonable agreement with the theoretical value from equation
(9.5).

The absorption of gold doped *p* type germanium is shown in
Figure 10.11. The long wavelength absorption edge in the 8 μ

146

region corresponds to transitions to the acceptor level 0·15 eV above the valence band. This edge disappears at room temperature, when there are large numbers of free holes and the acceptors are largely filled by electrons from the valence band, and in fact the absorption then *rises* considerably with increasing λ between 6 and 12 μ.

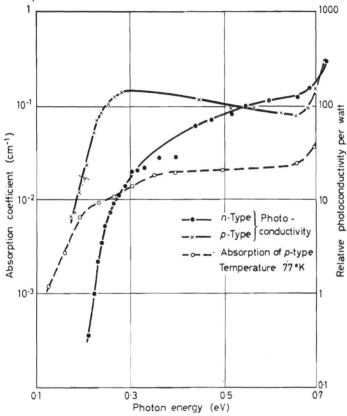

Figure 10.11. *Absorption and photosensitivity of gold doped germanium.*
(*Newman,* 1954; courtesy *American Physical Society*)

10.8.2 *Photoconductivity in doped germanium*

Photoconductivity arising from excitation of impurity levels in Ge has now been observed with a variety of impurities, particularly the noble and transition elements. As with silicon, it is necessary to operate at temperatures low enough to prevent thermal ionization of the appropriate level, and when impurities have multiple levels,

147

different spectral sensitivity curves are obtained with p and n type material.

One of the most interesting doping agents from the point of view of a practical, sensitive, i.r. detector is gold, as the activation energies are large enough for the detector to be operated at liquid air temperature. The spectral sensitivity curves of both p and n type material (from Newman, 1954) are shown in *Figure 10.11*. Because of the lower activation energy of the acceptor state near the valence band the p-type material is sensitive to longer wavelengths than the n-type. The sensitivity of the latter has fallen to half its value in the short wavelength region (~ 2 μ) at 3 μ, and to 10 per cent at 4·5 μ. The p-type response in *Figure 10.11* has fallen to half maximum at $\lambda_{1/2} = 5·5$ μ. Kaiser and Fan (1954) find $\lambda_{1/2} = 6·5$ μ.

For n type material, Woodbury and Tyler (1957) found a response time ~ 100 μsec at 77°K. These workers established the four energy levels of the gold centres given in *Table 10.2*.

No figures have been published for the sensitivity of gold doped germanium detectors, but it is expected to be high, since the responsivity is known to be high from the fact that most of the conductivity of the cooled specimens is photoconductivity produced by room temperature radiation. (Lasser *et al.* give N.E.P. $= 7 \times 10^{-11}$ watts at 77°K for area 2·25 mm².)

Copper doped germanium is a promising infra-red detector for much longer wavelengths. As shown by *Figure 10.10*, its quantum efficiency is roughly constant from 2 μ to 20 μ, and $\lambda_{1/2} = 27$ μ. A similar value has been given by Burstein *et al.* (1954). This detector requires cooling with liquid helium. Woodbury and Tyler (1957) found the time constant to be less than 3 μsec.

Zinc doped germanium has been studied by Burstein *et al.* (1954). The quantum efficiency, at helium temperatures, is nearly constant from 4 μ to 30 μ, and has fallen to 60 per cent of this value at 38 μ, the limit of the measurements. It is thus reasonable to put $\lambda_{1/2} = 39\text{--}40$ μ. This specimen had a time constant <10 μsec.

Photoconductivity of germanium doped with Mn, Fe, Co and Ni has been investigated by Newman and Tyler (1954) and Newman *et al.* (1956). Their results for Co and Fe doped p and n type material are shown in *Figure 10.12*. As expected from *Table 10.2*, the longer wavelength response is with p type for Co doping and n type for Fe doping. Ni doped p type material has a rather lower activation energy, and the photoconductivity is down to 10^{-3} of its 2 μ value at 0·2 eV. At 65°K, Mn doped Ge gives $\lambda_{1/2} = 5$ μ, so that the response of this material is almost as good as the gold doped, p type, germanium.

A theoretical analysis of the relative importance of various recombination mechanisms in determining the carrier lifetimes in doped Ge and Si has been given by Sclar and Burstein (1955). Generation–recombination noise has been observed by Hill and van Vliet (1958) and used to find lifetimes.

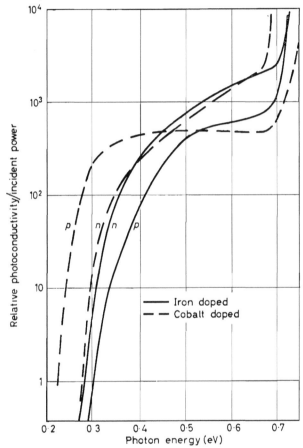

Figure 10.12. Photoconductivity of Fe *and* Co *doped germanium.*
(*After Newman et al.,* 1954, 1956)

The kinetics of photoconductivity in doped Ge are governed by the relations given for silicon in section 9.9. According to Brooks (1955) the density of states parameters for the conduction and valence bands are:

$$N_c = 0.41 \times 2(2\pi m k T/h^2)^{3/2}, \qquad N_v = 0.22 \times 2(2\pi m k T/h^2)^{3/2}$$

149

Thus, as for Si, there will be a lower free carrier concentration—for a given activation energy—with p type material.

10.9 GERMANIUM–SILICON ALLOYS

Germanium and silicon form solid solutions over the whole range of composition. It is possible to grow single crystals with any mixture ratio, although it is much more difficult to do so than with the elements. It has been pointed out by Parmenter (1955) that in such alloys the edges of the bands will be less well defined than in monatomic crystals.

The main technical interest in these alloys arises from the following possibilities: (1) a semi-conductor of adjustable energy gap, (2) devices with spatially varying energy gap (Kroemer, 1957), and (3) a semi-conductor of adjustable dielectric constant, and hence adjustable impurity activation energy. On the fundamental side, additional information about the energy band structure is obtainable from study of such alloys.

The energy gap does not vary uniformly with composition, but increases more rapidly as Si is added to Ge than for the converse, as shown by measurements of the thermal energy gap by Levitas et al. (1954) and of the optical energy gap by Johnson and Christian (1955), and Braunstein et al. (1958), whose results are shown in Figure 10.13.

It will be seen that the gap increases approximately up to linearly 15 mol. per cent Si. At this concentration there is a kink in the curve, and for higher concentrations there is a slower increase in gap until pure silicon is obtained.

Herman (1955a) has interpreted these results to mean that as silicon is first added to germanium, the lowest conduction level of the latter (i.e. in the [111] direction) moves steadily away from the valence band. At the other end of the range of compositions, addition of Ge slowly brings the [100] conduction band minimum of Si nearer to the valence band. At ∼15 per cent Si, the two conduction band minima are assumed to be at the same energy level. Cyclotron resonance measurements on Ge alloyed with small percentages of Si support this theory (Dresselhaus, Wagoner et al., 1955).

The availability of single crystals of Si/Ge alloys with dielectric constants between those of the pure elements indicates that impurity activation energies of doping elements will vary between those given in Tables 9.3 and 10.2. In particular, it should be possible to adjust one of the energy levels to lie near 0·1 eV, which is estimated to be the minimum energy gap for a satisfactory infra-red detector to operate at liquid nitrogen temperature. One obvious possibility is to reduce the ionization energy of In in Si (0·16 eV) by the addition of some Ge. Another, is to increase the energy of the

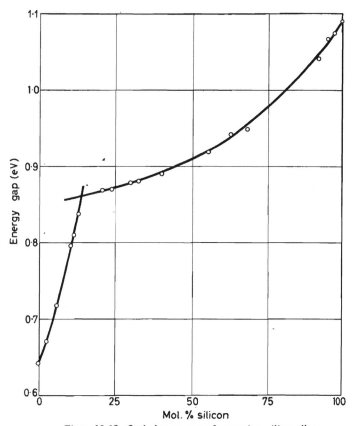

Figure 10.13. Optical energy gap of germanium–silicon alloys.
(*Braunstein* et al., 1958; courtesy *American Physical Society*)

Zn level in Ge by adding Si. At present it is not known how the dielectric constant varies with concentration, but it has been shown that the lattice constant varies linearly with composition over the whole range (Johnson and Christian, 1955).

CHAPTER 11

SELENIUM

11.1 INTRODUCTION

SELENIUM exists in various allotropic forms, all of which are semi-conductors (or near insulators) and which show photoconductivity. These forms are:—

(1) *Amorphous selenium.* The properties of this material seem to join those of liquid selenium without discontinuity. It is almost an insulator at room temperature.

(2) *Monoclinic crystalline red selenium.* This is also of very high resistivity. According to Weimer and Cope (1951) there are identifiable α and β forms.

(3) *Hexagonal, gray, selenium.* By contrast with the other forms this allotrope is a relatively good conductor and is known as 'metallic' selenium, although as its room temperature resistivity is $\sim 10^6$ Ω cm the term is rather an exaggeration.

Gray selenium has a regular crystal lattice built up of parallel spiral chains of atoms with covalent binding between atoms and relatively weak van der Waals forces between the chains. The lattice constants are $a = 4\cdot355$, $c/a = 1\cdot365$ (Straumaris, 1940).

The monoclinic forms consist of Se_8 rings or ordered Se_8 chains, whilst amorphous and liquid selenium are presumably made up of randomly oriented chain molecules. The amorphous form is produced by vacuum evaporation or super-cooling of liquid Se. It converts to the gray form very slowly at room temperature, but rapidly at $\sim 150°C$ (Keck, 1952).

The main commercial interest lies in photo-voltaic cells of 'metallic' selenium, but of recent years the most fundamental interest has been centred on amorphous selenium and most of this chapter will be devoted to this form.

11.2 ABSORPTION IN AMORPHOUS SELENIUM

11.21. *Short wavelength absorption*

Amorphous selenium absorbs strongly in the u.v. and short wavelength part of the visible spectrum, but becomes almost perfectly transparent at the red end of the visible region. Transmission measurements on thin amorphous films have been made by Gilleo (1951), Stuke (1953) and Hilsum (1956). Other results have been summarized by Moss (1952*a*). All the workers prepared their films by vacuum evaporation. Their results are plotted in *Figure 11.1*—the good agreement between them shows that the absorption coefficient in this highly absorbing region can be considered to

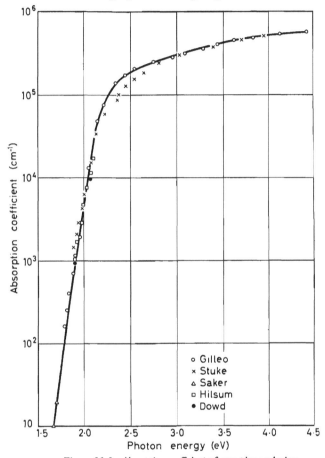

Figure 11.1. Absorption coefficient of amorphous selenium

be accurately known. It may be noted that on the semi-logarithmic plot used the absorption coefficient falls linearly in the neighbourhood of the absorption edge; i.e. it varies exponentially with photon energy over the extensive wavelength interval 0·55–0·8 μ. The rate of fall is a drop of absorption coefficient of e:1 for a change of photon energy of 0·067 eV, i.e. about $3kT$. This contrasts with

153

the slope of e:1 absorption change per kT quoted for several materials in section 3.2.1.

Also included in *Figure 11.1* are some results by Saker (1952) and Dowd (1951). Measurements at shorter wavelengths by Gilleo (1951) show that there is a broad maximum in K at 0·26 μ and in k near 0·3 μ.

There is some difficulty in deriving a specific value for the energy gap of selenium from absorption data. The results of Gebbie and Saker (1951) for example show that absorption extends to wavelengths at least as long as 1 μ (where $K = 2$ cm^{-1}) so that it seems inappropriate to try to discover the point where absorption ceases. It is preferable to use the definition of section 3.2.1, namely the position of greatest slope of a linear plot of K against $h\nu$. Using the data of *Figure 11.1* it is found that between photon energies of 2·1 and 2·4 eV the slope is steepest.

Stuke (1953) found that the reflectivity of amorphous selenium changed considerably over the energy range 1·9–2·3 eV. He preferred to identify the activation energy with the 'corner' on his graph at 2·3 eV. As far as optical data are concerned therefore the preferred value for the activation energy is 2·2–2·3 eV. This is considerably higher than the value of 1·86 eV deduced by Fochs (1956) from his powder reflection spectra. It is however in good agreement with values of thermal activation energy determined from resistance–temperature measurements on liquid selenium, namely 2·3 eV by Henkels (1950), 2·28 eV by Lizell (1952) and 2·31 eV by Henkels and Maczuk (1953).

11.2.2 Shift of the absorption edge

Spectral measurements of the absorption coefficient of liquid selenium were made by Saker (1952) for temperatures between 18°C and 414°C. There is a marked movement of the edge to longer wavelengths on heating. The shift of the absorption edge at a level of 20 cm^{-1} varies from 11×10^{-4} eV/°C for the lower temperature range to ∼22 $\times 10^{-4}$ eV/°C near 400°C. Hyman (1956b) obtained 18×10^{-4} eV for the 200–400°C range. Measurements on cooled films were made by Gilleo (1951) whose results at an absorption level of 700 cm^{-1} give a shift of 9×10^{-4} eV/°C between 200 and 300°K, falling to only $5·3 \times 10^{-4}$ eV/°C in the neighbourhood of 100°K. These results would thus appear to join smoothly on to those of Saker (1952). Hilsum (1956) made an accurate determination of the shift at room temperature by observing the change of transmission of sodium light with temperature for films of various thicknesses. These results gave the temperature variation of the absorption coefficient as $\mathrm{d}K/\mathrm{d}T = 190\ \mathrm{cm}^{-1}/°\mathrm{C}$, which from *Figure 11.1* is seen to correspond to $9\cdot7 \times 10^{-4}\ \mathrm{eV}/°\mathrm{C}$ at an absorption level $\sim 10^4\ \mathrm{cm}^{-1}$.

This behaviour is in contrast to most semi-conductors where the shift is considered temperature independent. Here it is roughly proportional to absolute temperature. Similar behaviour of the shift in diamond is discussed in section 8.1.

In section 3.3 it is shown that the temperature dependence of the absorption edge is the sum of two effects, the dilatation effect and the lattice broadening effect. Equation (3.20) shows that the dilatation effect is directly proportional to the thermal expansion coefficient. As this is extremely large for selenium $(56 \times 10^{-6}$ per °C according to Sato and Kaneko, 1949) it is reasonable to assume that the dilatation effect will dominate. Putting the value for the total shift in equation (3.20) gives

$$\pm C_e \pm C_h = 9\ \mathrm{eV} \text{ per unit dilatation}$$

This expression indicates quite moderate values for the Cs, comparable with estimates for InSb (see section 16.5.1) for example, thus showing that the large shift is mainly accounted for by the large expansion coefficient.

Now as shown by equation (3.19) the constants C_e and C_h have an important influence on the values of the carrier mobilities. Putting numerical values into equation (3.19) we obtain

$$C^2 \mu (m^*/m)^{5/2} = 1\cdot3 \times 10^{-8} c_{ll}$$

From the values of c_{ll} for Si, Ge and Te (Shockley, 1950) we can estimate that for Se it will be $\sim 0\cdot2 \times 10^{12}$ dynes/cm². Hence, if the deformation potential theory is applicable,

$$\mu (m^*/m)^{5/2} \doteqdot 2600/C^2$$

Taking $C \sim 9\ \mathrm{eV}$, $\mu (m^*/m)^{5/2} \sim 32$.

Now the mobilities in amorphous selenium have been measured by Spear (1957) who finds

$$\mu_h = 0\cdot15\ \mathrm{cm^2/V.sec}$$

$$\mu_e = 5 \times 10^{-3}\ \mathrm{cm^2/V.sec}$$

These values are of course extremely low compared with any other monatomic semi-conductor. When inserted in the above equation they yield the following values for the effective masses;

$$m_e \sim 33, \qquad m_h \sim 9$$

155

These values are very large, but are of the same order as the estimate of twenty free electron masses deduced by Schottky (1953) from electrical measurements on selenium.

A value of effective mass which should be accurate may be obtained from the Faraday effect, using equation (5.10). Becquerel (1877) observed that the specific Faraday rotation of selenium was 10·9 times that of CS_2 at $\lambda \sim 0.68$ μ. Hence $\theta = 99$ rds. per weber m^{-2}, and at this wavelength, $\lambda\, dn/d\lambda = 0.82$, so that

$$m^* = 2.5m$$

This latter value of effective mass is presumably much more realistic than that determined from the shift and mobility data, particularly since (as discussed in section 11.7) the mobility is probably determined mainly by intermolecular barriers.

11.2.3 *Long wavelength absorption*

Amorphous selenium has high transparency in the infra-red region. Between 2 and 4 μ Gebbie and Saker (1951) found that the absorption coefficient was ~ 0.5 cm^{-1}.

In the longer wavelength region Gebbie and Cannon (1952) observed well defined absorption bands at 13·5 and 20·5 μ. These workers attribute the bands to the Se_2 molecule by analogy with the 7·8 and 12 μ bands observed in plastic sulphur by Taylor and Rideal (1927). The wavelength ratio between both the long wavelength and short wavelength bands is 1·72 : 1, which is somewhat higher than the square root of the mass ratio of selenium to sulphur, indicating that the force constant of the Se–Se band is rather lower than that of the S–S band.

There is no observable free carrier absorption for wavelengths up to 25 μ—which is as indicated by equation (2.32) when the appropriate parameters are inserted.

Differences in the values of the dielectric constant and (refractive index)^2, described in the next section, indicate that considerable absorption must occur somewhere in the long wavelength infra-red region.

11.3 REFRACTIVE INDEX OF AMORPHOUS SELENIUM

A review of refractive index data for amorphous selenium has been given previously (Moss, 1952a). Since then comprehensive measurements of the temperature dependence of the refractive index of liquid (and super-cooled liquid) selenium have been reported by Saker (1952) whose results are summarized in *Figure 11.2.*

156

It will be seen that the temperature dependence of the index is large. As with the absorption edge shift, this is mainly the outcome of the high expansion coefficient. Saker (1952) found that the

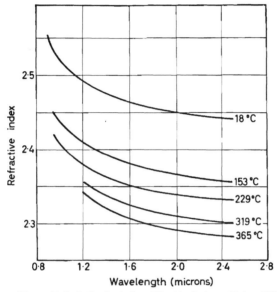

Figure 11.2. Refractive index of liquid selenium. (*Saker, 1952; courtesy Physical Society, London*)

temperature dependence of the polarizability per volume (for long wavelengths) was

$$\frac{d}{dT} \log \frac{(n^2 - 1)}{(n^2 + 2)} = -1.7 \times 10^{-4} \text{ per } °C$$

while the volume expansion coefficient over the same temperature range was given by

$$\frac{d}{dT} \log (\text{volume}) = 1.5 \times 10^{-4} \text{ per } °C$$

Hence the polarizability per atom of selenium is virtually temperature independent.

At wavelengths greater than 3 μ, Gebbie and Saker (1951) found that the refractive index became constant at 2·45, while Dowd (1951) obtained 2·46. Hence the non-dispersive infra-red dielectric constant is near 6·03. For low radio frequencies however, the dielectric constant is 6·3 (see Gebbie and Kiely, 1952).

In an attempt to discover in what region of the spectrum the change in these values occurs, Gebbie and Kiely (1952) made measurements at a wavelength of 3 cm, and obtained a dielectric constant of 5.97 ± 0.04, agreeing with the infra-red value. However, later measurements in the same laboratory—which are presumably more reliable—by Klinger and Saker (1953) show that at wavelengths of both 1.2 cm and 3 cm the value is 6.37. This value is substantially the same as the low frequency value, and the conclusion therefore is that the change occurs in the long wavelength infra-red region, and considerable absorption must thus occur somewhere in this part of the spectrum.

Part of the change in the dielectric constant could be due to the absorption band at 20 μ. However, integration over this band according to equation (2.29) shows that its contribution to the dielectric constant is only 0.001, so that to explain observed change a much stronger absorption band is needed. The requirement is approximately that

$$\int K \, d\lambda \sim 1$$

over the band.

Eckart and Rabenhorst (1957) found that the dielectric constant of hexagonal selenium decreased on heating, the average change being $d\varepsilon/dT = -18 \times 10^{-4}$. For amorphous material, ε decreased gradually with rising temperature up to 75°C, but with further heating it increased rapidly.

The measurements of refractive index available do not extend to sufficiently short wavelengths for equation (2.26) to be used to calculate the value of dielectric constant in the near infra-red in order to compare it with the observed value. An Argand diagram of the real and imaginary parts of the dielectric constant for amorphous selenium, using data of Meier (1910) and Wood (1902) for refractive index and Gilleo (1951) for absorption, is not a circle as expected from the theory of section 2.2 for a classical oscillator.

11.4 Photoconductivity in amorphous selenium

It was thought for many years that in contrast to the behaviour of the metallic form of selenium, the amorphous form did not show photoconductivity. However independent measurements by the author and by Weimer in 1950 and Gilleo in 1951 established that the material was photosensitive. The photocurrents are normally very small but by virtue of the very high dark resistivity they can be many times the dark current. By use of sandwich type cells only 5 or 10 μ thick, the application of 30 V gives a sufficiently high field to produce a sustained photocurrent of 1 $\mu\alpha$ per cm^2 (Weimer and Cope, 1951).

Some spectral sensitivity curves are shown in *Figure 11.3*. It will be seen that for thicker layers maximum sensitivity occurs near 0·45 μ, the quantum yield being near unity for wavelengths shorter than this. The data of Weimer and Cope (1951) and that of Moss (1952a) both have $\lambda_{1/2} \doteqdot 0\cdot51$ μ, and a response down to 2 per cent of maximum at 0·59 μ. The data of Gilleo (1951) are biased towards shorter wavelengths—probably because his layer was only 0·57 μ thick, in contrast with the others which were both $\sim$5 μ thick.

Also shown in *Figure 11.3* (curves *a* and *b*) are the computed absorption curve for the 5 μ layer and the actual measured absorption for the 0·57 μ layer. In each case it will be noted that the photoconductivity falls off at much shorter wavelengths than the absorption. The difference in photon energy at corresponding absorption and photoresponse levels are given in *Table 11.1*.

Table 11.1

Thickness	Absorption	Photoresponse	Energy Difference
5 μ	90% at 0·615 μ	90% at 0·475 μ	0·6 eV
5 μ	69% at 0·64 μ	69% at 0·495 μ	0·56 eV
0·57 μ	90% at 0·58 μ	90% at $\sim$0·43 μ	$\sim$0·75 eV

Weimer and Cope (1951) observed that the photocurrent is carried predominantly by positive holes which they found to have a range $\sim$10^{-3} cm in a 5 × 10^4 V/cm field. Measurements with pulses of light of 50 μsec duration showed that the response time was less than this value. From the expression, range $= \mu F\tau$, using the above values we find

$$\mu_h\tau \sim 2 \times 10^{-8}$$

Recent measurements of transit times of charge carriers produced by very short duration electron or light pulses by Spear (1957) give consistent values for the mobilities. At room temperature they are

$$\mu_e = 5\cdot2 \times 10^{-3}, \qquad \mu_h \doteqdot 0\cdot15 \text{ cm}^2/\text{V.sec}$$

In conjunction with the value for the $\mu_h\tau$ product given above we obtain an estimate of the lifetime

$$\tau \sim 10^{-7} \text{ sec}$$

Spear (1957) determined the temperature variation of mobility. He found exponential laws for both types of carriers, the results being

$$\mu_e \propto \exp\,(-0{\cdot}25/kT), \qquad \mu_h \propto \exp\,(-0{\cdot}16/kT)$$

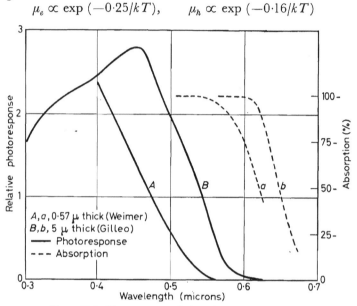

Figure 11.3. Photoresponse and absorption in amorphous Se

11.5 HEXAGONAL SELENIUM

Gilleo (1951), Stuke (1953) and Stegman (1957) have studied the change in the absorption spectrum on converting an evaporated film of amorphous selenium to the metallic form by heating. All found that the absorption was increased in the region of 0·6 μ. For photon energies between 2 and 2·5 eV Stuke's results show that the absorption curve is displaced approximately 0·25 eV to longer wavelengths on conversion; Gilleo's results show a shift ∼0·4 eV. Both workers find that the absorption edge of hexagonal Se moves to shorter wavelengths on cooling, the magnitude being 4×10^{-4} eV/°C at an absorption level of 10^5 cm^{-1}.

By subtracting the absorption of the amorphous material from that of the metallic, Stuke showed that the additional absorption was a band extending from ∼1·8 eV to ∼2·7 eV with its peak near 2·2 eV (at room temperature). This band coincides roughly with the spectral sensitivity curve of hexagonal selenium given by Gilleo (1951), whose results show that for a 0·65 μ thick specimen the

160

response has fallen to $1/e$ at $1\cdot 8$ eV. From Stuke's absorption data the corresponding absorption coefficient, i.e. $1\cdot 5 \times 10^4$ cm^{-1}, occurs also at $1\cdot 8$ eV, so that unlike the amorphous material there is no significant energy difference between the absorption edge and the fall-off of photoconductivity.

Choyke and Patrick (1957b) have tried to use photo-effects in metallic selenium to estimate absorption coefficients. Unfortunately they used a photo-voltaic method where the illumination had to pass through electrodes of gold and CdS before reaching the Se layers, rather than a photoconductive method where a single crystal could be illuminated between the electrodes. As the effective specimen thicknesses were not known, no absolute values of absorption constant could be obtained and it is doubtful if the attempt to fit the data to the theoretical expression for indirect transitions (equation 3.11) is justified. In order to obtain reasonable agreement with this equation it was found necessary to use a different value of phonon energy at each measuring temperature. The estimated energy gap from the results was $1\cdot 8$ eV, which agrees with typical $\lambda_{1/2}$ values from photo-voltaic spectral sensitivity curves, but exceeds the corresponding values from photoconductive response curves where $\lambda_{1/2}$ is usually near $0\cdot 8$ μ (see Moss, 1952a).

Blet (1956) has reported measurements of spectral sensitivity of a selenium photocell where two bands of sensitivity appear; a main band peaking at $0\cdot 55$ μ and a subsidiary one peaking at $\sim$$0\cdot 7$ μ. These two bands could be due to amorphous and hexagonal allotropes respectively. The interesting feature of Blet's results is that at $100°$K he finds that the short wavelength band disappears, while the $0\cdot 7$ μ band is relatively unaffected by cooling.

Dowd (1951) observed that there was considerable absorption in metallic Se crystals in the wavelength range $0\cdot 8$–2 μ, where he found $K \sim 550$ cm^{-1} in some specimens. Choyke and Patrick (1957b) question these results because the photoresponse does not continue into this region, and suggest that reflectivity corrections were inadequate. As Dowd did in fact make a reflectivity correction, and as for one of his specimens an error of 2:1 would only decrease K to 400 cm^{-1} or increase it to 700 cm^{-1}, any inaccuracy much worse than this seems unlikely. Also, if the absorption is due to free carriers or scattering at imperfections within the crystals, there would not of course be any corresponding photoresponse.

Soft x-ray measurements by Givens et al. (1955) show a doubly peaked curve in the region of 220 Å. The energy separation of these peaks is approximately $1\cdot 2$ eV in both vitreous and metallic specimens.

11.6 REFLECTIVITY OF SELENIUM

Reflectivity of both amorphous and metallic selenium has been measured by Stuke (1953) for photon energies between 1 and 4 eV, and by Kandare (1957) for energies up to 11 eV. Their results are plotted in *Figure 11.4*.

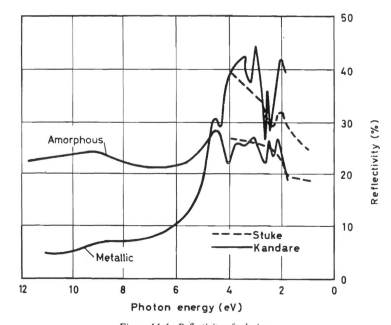

Figure 11.4. Reflectivity of selenium

The two sets of results are in general agreement if it is assumed that the rapid variations shown by Kandare are caused by interference effects. The reflectivity of the metallic material falls greatly between 4 and 6 eV and is only about 7 per cent at 7 eV. This behaviour would seem quite reasonable on the grounds that the main absorption could well drop in this region. For the amorphous material however, the reflectivity remains greater than 23 per cent for photon energies up to nearly 12 eV.

Pribytkova (1957) has used reflection measurements, in the visible region, to determine n and k.

From powder reflection spectra, Fochs (1956) estimates the energy gap of metallic Se to be 1·74 eV.

162

11.7 DISCUSSION

Compared with well known crystalline semi-conductors such as germanium, the phenomena of photoexcitation and conduction in selenium are still not well understood. The properties of selenium which need explanation are:—

(1) Difference in photon energy for absorption and photo-conduction.

(2) Low mobility and its exponential temperature dependence.

(3) Non-linear current voltage relationships. (Weimer and Cope, 1951; Gobrecht and Hamisch, 1957.)

(4) Carrier concentration independent of temperature in hexagonal selenium, with $\sim 10^{14}$ carriers/cm^3 (Plessner, 1951; Kozyrev, 1957.)

(5) Marked frequency dependence of conductivity (Henkels, 1951; Henkels and Maczuk, 1954). At 200 Mc/s the room temperature resistivity is ~ 100 times less than the d.c. value.

It is suggested that all these characteristics arise from potential barriers within the material which result from the basic structure of selenium. It is assumed that within individual chains or rings of selenium atoms carriers may move freely, but that to travel to neighbouring chains or rings it is necessary to cross potential barriers. Such proposals have also been made by Hyman (1956a) and Billig (1952).

If we postulate that these barriers are entirely responsible for the difference in photon energy between equivalent photo-conduction and absorption levels in amorphous selenium—i.e. the limit of absorption corresponds to the lowest energy required to excite carriers across the energy gap, but the limit of photoconductivity corresponds to raising carriers to sufficiently high energy levels that they can cross the inter-molecular barriers, then the barrier height is ~ 0.6 eV.

We would therefore expect that if the mobility were measured using low-energy radiation it would vary exponentially with this activation energy. In fact the results of Weimer and Cope (1951) using excitation by visible light show an approximately exponential dependence with an activation energy of 0.8 eV—which is reasonable agreement. In his experiments, Spear (1957) used electron bombardment (where ~ 20 eV was given up per electron-hole pair produced) or light from a spark gap which would be rich in ultra-violet radiation. The use of such high energy excitation found by this latter worker may therefore be the reason that the activation energies of mobility were correspondingly lower.

163

In liquid selenium at temperatures $\sim 300°C$ Henkels and Maczuk (1953) find high mobilities of ~ 1000 cm²/V.sec so that apparently when the molecules are freely mobile in the liquid state the intermolecular barriers disappear.

In metallic selenium the temperature dependence of the mobility of the dark current carriers has a lower exponent, namely

$$\mu_h \propto \exp\left(-0.13/kT\right)$$

This is in accord with the observations of Gilleo (1951) that in this allotrope there is no significant difference between the absorption and photoconductive spectra. In fact the photocurrent curve shows a peak in the region of the absorption edge, the response falling again at short wavelengths—probably as a consequence of surface recombination as treated theoretically in section 4.2.3.

If barriers are much larger in amorphous selenium than in hexagonal selenium, then cooling would be expected to reduce photoconductivity in the former much more than in the latter. This may be the explanation of the observation of Blet (1956) described in section 11.5.

11.8 MIXTURES OF SELENIUM WITH
TELLURIUM OR SULPHUR

Selenium–tellurium mixtures are isomorphous in all concentrations and so the properties can be studied as the relative proportions are changed progressively. So far data are available only for alloys with approximately 0–10 per cent Se or 0–10 per cent Te. However even small additions of the other component cause marked changes.

Keck (1952) has shown that the resistivity of selenium falls by about ten times when 8 per cent Te is added, and that at the same time the threshold wavelength moves from $\lambda_{1/2} = 0.5$ μ to $\lambda_{1/2} = 0.67$ μ.

Nussbaum (1954) has found that addition of 11 per cent Se increases the resistivity of Te from 0.54 Ω cm to 9.5 Ω cm, while the same addition was found by Loferski (1954) to reduce $\lambda_{1/2}$ from 4.02 to 3.0 μ.

Although the range of mixtures so far observed is small, *Figure 11.5* indicates that $\log_{10}$ (resistivity) varies approximately linearly with tellurium concentration. The figure also shows that the dependence of $\lambda_{1/2}$ is much larger at the 100 per cent Se end of the curve, which is in accord with the knowledge that the temperature and dilatation dependence of the energy gap is much larger for Se than for Te.

Selenium and sulphur are also miscible in all proportions.

Transmission measurements on liquid mixtures of the two have been made by Hyman (1956*b*) in the temperature region 200–400°C. His results for the position of absorption edge (taken as 2 cm^{-1} absorption coefficient at 210°C) and its temperature dependence are given in *Table 11.2*.

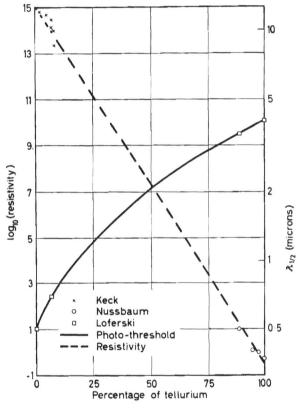

Figure 11.5. Photoconductivity and resistivity of selenium with tellurium additions

The shift thus remains characteristic of amorphous Se alone for proportions of S of at least 84 per cent, in spite of the fact that the optical energy gap is varying continuously with increasing Se concentration. It may be noted however, that the change in energy gap with concentration is much greater for 84–100 per cent S than for <84 per cent S.

A possible explanation is that the edge is in fact determined by

165

the selenium (and hence the shift is characteristic of selenium), the apparent edge variation with concentration being due to the change in *intensity* of the selenium absorption. From the curves of Hyman (1956*b*) for concentrations of 16 and 76 per cent Se,

Table 11.2. Absorption Data for Se/S Mixtures

Composition	Edge (at 210°C)	Shift (eV/°C $\times$ 10^{-4})
100% Se	1·15 eV	18·5
24% S	1·29 eV	18·2
59% S	1·45 eV	18·8
84% S	1·67 eV	18·1
100% S	2·13 eV	29·2

we may determine the absorption edges at a level proportional to the concentration. The values at 210°C are:

16 per cent Se, photon energy at 1 cm^{-1} $\quad = 1\cdot56$ eV
76 per cent Se, photon energy at 76/16 cm^{-1} $= 1\cdot44$ eV

Thus, on this basis, the additional 60 per cent Se moves the edge only 0·12 eV compared with the shift of 0·6 eV caused by the addition of the first 16 per cent Se to the sulphur. To a first approximation therefore we may say that the position of the absorption edge at a given oscillator strength, as well as the shift of the edge, are determined by the selenium.

CHAPTER 12

TELLURIUM

12.1 INTRODUCTION

TELLURIUM crystals are built up of spiral chains of atoms of inter-atomic spacing 2·86 Å, the distance of closest approach of atoms in neighbouring chains being 3·45 Å. In consequence the crystals are highly anisotropic, the conductivity differing by a factor of two in directions parallel to and perpendicular to the crystal axis. The electrical properties are discussed by Fukuroi (1949, 1950), Moss (1952a), Nussbaum (1954), Long (1956), Tanuma (1954) and Fischer *et al.* (1957). Unlike selenium, tellurium appears to exist in only one form.

12.2 ABSORPTION

It was shown by the author (Moss, 1952b) that bulk tellurium was transparent to the longer infra-red wavelengths, the absorption edge being at about 3·5 μ.

Comprehensive measurements have been carried out in the neighbourhood of the absorption edge by Loferski (1954). Working with single crystals and polarized radiation he established that the absorption edge moved significantly when the radiation was changed from its electric vector parallel to the *c* axis to the condition of electric vector perpendicular to the *c* axis. The absorption coefficient calculated from Loferski's transmission data is plotted in *Figure 12.1*.

Transmission measurements on films evaporated onto plates of artificial sapphire (Moss, 1952b) enabled the absorption curve to be plotted at high levels. Such data are also shown in *Figure 12.1*. Loferski takes the absorption edges to be defined by the condition that the transmission is 5 per cent of its value in the transparent region, and obtains:

$$h\nu_{\parallel} = 0.374 \text{ eV}, \qquad h\nu_{\perp} = 0.324 \text{ eV}$$

According to Dresselhaus (1957), theoretical treatment of the problem of absorption in anisotropic crystals shows that for indirect transitions the shape of the absorption edges would be the same for both allowed and forbidden transitions, whereas for direct transitions the laws should be different, namely $K \propto (h\nu - E)^{1/2}$ for

167

allowed, and $K \propto (h\nu - E)^{3/2}$ for forbidden transitions. Dresselhaus also shows that for band maxima and minima of likely types, polarizations parallel to the c axis should be allowed and those normal to the c axis forbidden. These theories are clearly quali-

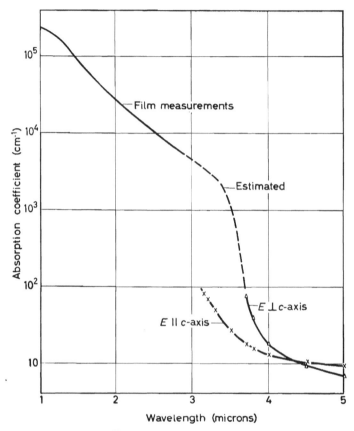

Figure 12.1. Absorption in tellurium

tatively in accord with the experimental data obtained on crystals shown in *Figure 12.1*. For absorption coefficients >25 cm^{-1} the data can be fitted reasonably well to the expressions

$$K_\parallel \propto (h\nu - 0.36)^{1/2}, \qquad K_\perp \propto (h\nu - 0.31)^{3/2}$$

The high K data of *Figure 12.1* indicate that the absorption at short wavelengths follows roughly the law $K \propto (\nu - \nu_0)^{1.5}$.

168

12.3 TEMPERATURE AND PRESSURE DEPENDENCE
OF ABSORPTION

Loferski (1954) carried out transmission measurements using polarized radiation at different temperatures. For the shorter wavelength edge, i.e. for the electric vector parallel to the c axis, his published curves show that for a temperature change from $-195°C$ to $129°C$ the wavelength of the edge (taken here as the point where the transmission is 5 per cent of its value in the non-absorbing region) increases by 0.24 μ. This therefore corresponds to a shift of -0.7×10^{-4} eV/°C. (Due presumably to an arithmetical error Loferski quotes the shift as 0.2 to 0.5×10^{-4} eV/°C.) For the E vector perpendicular to the c axis the shift was similar. As the shifts were near the limit of resolution of the equipment, no great accuracy is claimed for the results.

As discussed more fully later, Loferski's measurements of photoconductivity on single crystals showed two pronounced peaks whose wavelengths corresponded to the absorption edges. The short wavelength peak was observed to move on average by -7×10^{-5} eV/°C. This is in precise agreement with the absorption shift calculated above, although again high experimental accuracy was not claimed.

Neuringer (1955) has observed the shift of the absorption edge with pressure. Again polarized radiation was used, and for the E vector parallel to the c axis the movement was -1.9×10^{-5} eV/°atmos. So far a precise determination has not been reported for the other edge, but qualitatively it moves in the same direction.

This figure is rather larger than that deduced by Bardeen (1949) from Brigman's (1938) conductivity data, which for the range 0–20,000 atmospheres give -1.3×10^{-5} eV/atmos. Brigman's data are not representative of intrinsic Te at low pressures and the indication is that the change per atmosphere would be larger for the low pressure region where the optical shift measurements were carried out. If plotted against dilatation the activation energy varies linearly at high pressures and gives a zero pressure intercept at the accepted value of activation energy (Moss, 1952a), so that it can safely be assumed that at low pressures the shift is 3.5 eV per unit dilatation. The low pressure compressibility is $\chi = -5.1 \times 10^{-6}$ per Kg/cm^{-2} or -5.2×10^{-6} per atmos., so that the pressure shift becomes $dE/dP = -1.8 \times 10^{-5}$ eV/atmos. in excellent agreement with the optical determination.

An independent estimate of the shift with dilatation can be obtained from the measurements of Nussbaum (1954) on Te–Se alloys. These results show that the absorption edge moves to shorter wavelengths at a rate of 0·007 eV per 1 per cent addition of Se. Measurements by Grison (1951) on the lattice constant of Se–Te alloys enable us to calculate the dilatation. The value is $dV/V = -1·9 \times 10^{-3}$ per 1 per cent Se. Hence $dE/(dV/V) = 3·7$ eV per unit dilatation.

Theoretical estimates of the dilatation effect by Bardeen and Shockley (1950) and Fukuroi (1951) give values $\sim$4 eV per unit dilatation—in reasonable accord with the above experimental values.

As the thermal expansion coefficient is $\Delta V/V\Delta T = 53 \times 10^{-6}$, the dilatation should give a shift $+1·9 \times 10^{-4}$ eV/°C. In view of the observed temperature shift of the edge of -7×10^{-5} eV/°C, there remains a net shift of $-2·6 \times 10^{-4}$ eV/°C to ascribe to the lattice interaction term. Theoretical estimations of this term are given by Tanuma (1954) as $7 \pm 3 \times 10^{-5}$ eV/°C. As, however, the theory is greatly over simplified, good agreement could not be expected. Furthermore it should be noted that the photo-conductive threshold wavelength ($\lambda_{1/2}$ value) was found to have a small positive shift, both for films (Moss, 1952a) and for single crystals (Loferski, 1954), although for the latter the photoconductive *peaks* had a negative shift.

12.4 REFRACTIVE INDEX

Refractive index measurements have been made on tellurium in the visible region by van Dyke (1922) using reflection techniques. He found values in the region of 2·5–3·0. For wavelengths beyond the absorption edge the index of evaporated films was determined by Moss (1952a) interferometrically. These results showed that in the long wavelength non-dispersive region (i.e. $\lambda > 7 \mu$) the index had the extremely high value of approximately 5. The measurements were made with low order fringes so that the assignment of the fringe numbers was unambiguous. More recently Hartig and Loferski (1954) have made measurements on single crystals, using polarized radiation. They found that the indices were markedly different, depending on the orientation of the plane of polarization relative to the c axis. The prism angle used was approximately 5°, and for convenience of adjustment measurements were not made at minimum deviation but with the incident beam perpendicular to the first prism face and the deviation measured by simply locating the emergent beam. For this condition

$$n = \sin(\alpha + \Delta)/\sin \alpha$$

where α is the prism angle and Δ the deviation. The results obtained are shown in *Figure 12.2*. The dispersion is seen to increase as the wavelength approaches the absorption edges. For wavelengths beyond 8 μ, Caldwell's results quoted by Fan (1956) show that the dispersion is small.

Hartig and Loferski suggest that the weighted mean of their values, namely

$$\bar{n} = (2n_\perp + n_\parallel)/3$$

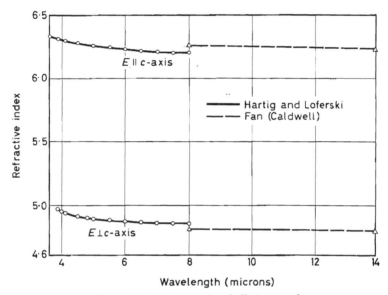

Figure 12.2. *Refractive index of tellurium crystals*

should give the average index observed in films if these are assumed to consist of randomly oriented microcrystals. This would give a value of $\bar{n} = 5 \cdot 3$ at 7 μ for example, where the film measurements gave $n = 5 \cdot 1$. This agreement is probably as good as can be expected in view of the uncertainties in the interferometric results. If, however, the evaporated layers grow crystals preferentially, for example with the c-axis parallel to the substrate, then radiation passed through the layers would always have the E vector perpendicular to the c-axis (even in unpolarized radiation) and the refractive index would tend towards the lower $(n_\perp)$ values. Such preferential growth has been reported for crystallized Se layers (see Barnard, 1930).

171

12.5 Long wavelength absorption

Measurements from the absorption edge to 25 μ are reported by Fan (1956) for both orientations of polarized radiation. The results are shown in *Figure 12.3* for various temperatures.

At the longest wavelengths (i.e. 20–25 μ) and the higher temperatures, both orientations show approximately (wavelength)^2 dependence of absorption. In this range therefore, equation (2.30*b*)

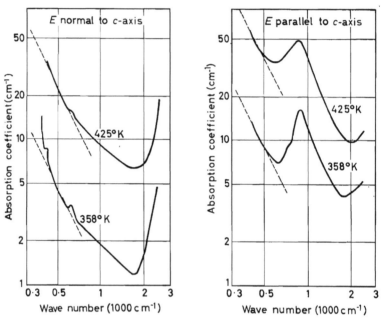

Figure 12.3. *Long wavelength absorption in tellurium.* (*Fan*, 1956; courtesy *Physical Society, London*)

for the free carrier absorption might be expected to apply. For 358°K, the results of Tanuma (1954) give

$$n_i \doteq 4 \times 10^{16}, \qquad \mu_n = 1100, \qquad \mu_p = 700 \text{ cm}^2/\text{V.sec}$$

Hence at 20 μ we estimate on the basis of unity effective masses, $K = 0.4$ cm^{-1}. This is a factor of 10–20 below the observed absorption. As $K \propto 1/(m^*)^2$, the discrepancy could well be explained by effective masses of about one quarter of the free electron mass.

172

As the hole and electron mobilities are roughly the same the $m^2\mu$ terms in equation (2.35) should be approximately equal, i.e.

$$m_n{}^2\mu_n = m_p{}^2\mu_p$$

Hence under given conditions equation (2.35) gives $K \propto 1/\mu$, and thus

$$K_\perp/K_\parallel = \mu_\parallel/\mu_\perp, \quad \text{or} \quad K_\perp/K_\parallel = \rho_\perp/\rho_\parallel$$

From Bottom (1948) the resistance ratio is 1·9:1. From *Figure 12.3* the absorption ratio in the 20–25 μ region at 358°K is 1·8 to 2·0:1, so that within the limits of the experimental errors precise agreement is obtained.

Figure 12.3 also shows a pronounced band peaking at 11 μ when the polarization has its E vector parallel to the c axis, but no corresponding absorption occurs for E perpendicular to the axis. Fan (1956) reports that results on samples with different hole concentrations show that the intensity of this band is proportional to the hole concentration. This indicates that there is another band about 0·1 eV below the top of the valence band, the holes present at the top of the valence band giving rise to optical transitions from the lower band. The fact that the band does not appear for the E vector perpendicular to the c axis indicates that for this case the transition is forbidden by the selection rules.

Small absorption bands can be observed at 14·8 μ and 24·4 μ with either direction of polarization if they are not masked by free carrier absorption. These bands are unrelated to the carrier concentration and Fan (1956) concludes that they are lattice absorption bands. Godefroy (1956) quotes an impurity activation energy—for holes—of 0·04 eV. Fan's measurements do not quite extend to sufficiently long wavelengths to cover this energy region.

12.6 BAND THEORY

Callen (1954) has attempted to explain the dependence of absorption edge on the plane of polarization by an analysis of the band structure based on the symmetry properties of a 'drastically simplified' tetragonal model of the crystal structure. He concludes that the valence band is derived from atomic $4p$ states whereas the conduction band derives from atomic d states. He estimates that the d levels will give overlapping bands, and shows that the selection rules indicate that electrons from the valence band will be excited into entirely different d bands by the two polarizations of the radiation, naturally implying two different absorption edges.

Soft x-ray absorption measurements by Woodruff and Givens

(1955), and Givens *et al.* (1955) show a marked doublet with a separation between the peaks of 1·6 eV. These two peaks should represent the distribution of states in the conduction band.

This x-ray evidence, and the long wavelength absorption results of Fan (1956), which have been obtained since Callen's band structure proposals were put forward do not fit well into his picture. Clearly the long wavelength absorption data indicate the presence of structure in the valence band, with a second band edge $\sim 0·1$ eV below the top of the band. As the valence band is formed from a degenerate p level, the two bands could well arise from splitting of this level—Dresselhaus (1957) states that the spin orbit splitting should be large in Te. Transitions from these split levels to the lowest level of the conduction band would then be the explanation of the dependence of the absorption edge on polarization.

This suggestion is in agreement with the later band theory of Reitz (1957). He finds that the atomic $5p$ levels split into three groups of three bands, with relatively small separation between the three bands in a group. The middle group will form the valence band, whilst the bottom of the conduction band is formed either by the lowest (p^-) level of the upper p group or by the lowest d level. From the form of the matrix elements for direct optical transitions this worker shows that for the $p^- \rightarrow p^-$ transition the matrix element vanishes for polarization perpendicular to the c axis, while for transitions from the other two levels to p^-, it vanishes for polarization parallel to the axis. A similar conclusion is reached for transitions from the three middle p levels to the lowest d level, and for both $p \rightarrow p$ and $p \rightarrow d$ transitions it is shown that for radiation perpendicular to the c axis the absorption edge will be at the longer wavelength, in accord with *Figure 12.1*.

Clearly the difference in the absorption edges is not uniquely defined but is a function of the absorption level at which it is measured (see *Figure 12.1*). Extrapolation of the edges would show that a 0·1 eV separation (as indicated by the long wavelength absorption band) would occur for $k \sim 300$ cm^{-1}, and such a separation could arise from the valence band structure suggested. The x-ray data can then be interpreted as due to two conduction bands with a separation $\sim 1·6$ eV, which is too large to show up in the available optical absorption spectra. Further evidence of the presence of structure in the valence band is provided by the Nernst effect determinations of Mochan (1956), and by the low temperature Hall and magnetoresistance measurements of Shalyt (1956), which show the presence of two types of holes with mobilities of ~ 500 and 10,000 cm^2/V.sec.

174

It is concluded by Dresselhaus (1957) that a pronounced difference in refractive indices for the two polarizations—as is found in Te—implies a band structure where conduction band minimum and valence band maximum occur at the same point in the zone, with the transition for one polarization allowed and the other forbidden. This is confirmed by Reitz (1957) who finds that both extrema occur at the edge of the zone.

It has been suggested by Mooser and Pearson (1956) that the main energy gap in the Te band structure is bridged by a band of low state density which would give rise to a small, temperature-independent, density of free charge carriers. At present there seems to be little evidence to support this hypothesis.

12.7 PHOTOELECTRIC EFFECTS

Photoconductivity in tellurium was first observed in films (Moss, 1949a) and detailed descriptions of results obtained with evaporated layers are given by Moss (1952a).

12.7.1 *Spectral distribution of photosensitivity*

Little or no sensitivity was observed until the layers were cooled, and in general only when cooled to liquid air temperatures was the sensitivity sufficient to carry out spectral measurements. The layers used were very thin and the response curves tended to follow the pattern of the absorption coefficient shown in *Figure 12.1*, so that the sensitivity had fallen considerably before the true absorption edge was reached. This effect was particularly marked at higher temperatures, and as a result determination of the cut-off $(\lambda_{1/2})$ wavelength in these cases was very difficult.

At 90°K the average value obtained was $\lambda_{1/2} = 3\cdot38\ \mu$ corresponding to 0·37 eV. Although the available temperature range was very restricted and the interpretation of the $\lambda_{1/2}$ points somewhat arbitrary, it was estimated that the effect temperature was to give a positive shift $dE/d\tau \sim 10^{-4}$ eV/°C.

Photoconductive measurements on single crystals have been reported by Loferski (1954). Again cooling was necessary. Polarized light was used, and results for the two positions of the plane of polarization are shown in *Figure 12.4*. For both orientations the threshold wavelength is at $\lambda_{1/2} = 4\cdot02\ \mu$, and the effect of temperature on this threshold is to give a positive shift $dE/dT \sim 10^{-4}$ eV/°C, similar to results on evaporated layers. However a significant feature of *Figure 12.4* is the presence of two peaks on the spectral sensitivity curves. The fact that the relative intensities of these peaks depend on the polarization show that they are related to the two absorption edges. Presumably the peaks are a consequence of surface recombination as outlined in section 4.2.3 and correspond to positions on the absorption edges where the absorption coefficient is about the inverse of the carrier diffusion length. Taking a lifetime of 50 μsec, (Redfield, 1955a) and an ambipolar mobility ~1000 cm²/V.sec, (Nussbaum, 1953) the intrinsic diffusion length at 100°K will be $L_i \sim 10^{-2}$ cm, indicating

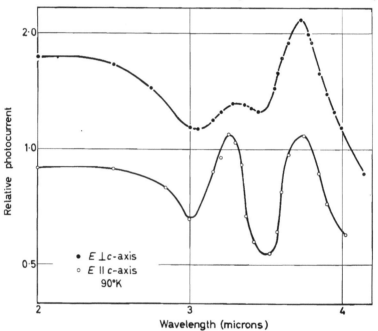

Figure 12.4. Spectral sensitivity of tellurium crystal. (Loferski, 1954; courtesy American Physical Society)

$K \sim 100$ cm⁻¹. With allowance for the temperature shift of the absorption edges, this figure corresponds fairly well with the absorption coefficients at the wavelengths of the peaks of the spectral sensitivity curves.

12.7.2 Intensity dependence of the photocurrent

It was found by Moss (1952a) that if a cooled tellurium layer was irradiated by relatively short wavelength illumination from a tungsten lamp, the photocurrent increased linearly with intensity at low levels of irradiation, but that at high levels the signal increased only as (intensity)$^{1/2}$. Similar behaviour has been found for single crystals by Loferski (1954) and Redfield (1955a) using 3 μ radiation, but the response was linear for the same photon intensity if 4 μ radiation was used (see *Figure 12.5*). This behaviour is as expected on the basis of simple recombination theory, (Moss, 1952a and Redfield, 1956). There is no evidence of an (intensity)$^{1/3}$ law at the highest intensities measured, as would be expected from the theory of Appendix C.

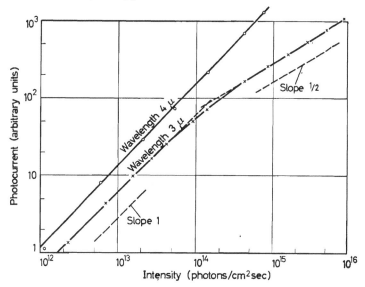

Figure 12.5. Photocurrent v light intensity in tellurium.
(Redfield, 1955b; courtesy University of Pennsylvania and U.S.
Office of Naval Research)

12.7.3 Carrier lifetimes

At 100°K the carrier lifetime in single crystals has been estimated to be ∼50 μsec (Redfield, 1956) from the magnitude of the photo-current, and 20 μsec from the decay time following a pulse of illumination from a spark gap (Redfield, 1955a). Drastic variations in surface treatment made no significant change in the photocurrent and hence, it is concluded, did not materially affect the lifetime. For layers, values ∼400 μsec were obtained at liquid air temperatures (Moss, 1952a).

For single crystals the photosensitivity was 10^4 times less at room temperature than at liquid air temperatures. For layers the corresponding figure was 10^5 times. Attributing the whole of the change to variation in lifetime, both layers and single crystals indicate a room temperature carrier lifetime $\sim 10^{-8}$ secs. Measurements of lifetime have been made by de Carvalho (1957) using the P.E.M. effect. He finds $\tau \sim 10^{-8}$ sec at room temperature and that cooling causes a rapid increase in the lifetime. Near room temperature, where the change is most rapid, $\tau \propto \exp E/kT$, with $E \doteq 0.30$ eV.

Redfield (1955a) concludes that direct electron-hole recombination is the predominant recombination mechanism at liquid air temperature. If this suggestion means radiative recombination, it is at variance with the estimates of radiation lifetime under such conditions. From equations (6.3) and (6.8), using refractive index data from *Figure 12.2* and absorption as estimated by the dotted line of *Figure 12.1*, it is found that at 300°K $\tau_R \sim 30$ μsec, with an error probably <3:1. At 100°K, making reasonable extrapolations of absorption and intrinsic carrier concentration, a crude value of $\tau_R \sim 1/3$ second is obtained, with a possible inaccuracy of perhaps 20:1. At both temperatures therefore a comparison with experimental values makes it appear that direct radiative recombination must be a very minor contributor. Redfield (1955a) searched for recombination radiation but was unable to detect any, possibly because of experimental limitations.

From his measured dependence of lifetime on temperature, de Carvalho (1957) suggests that the recombination process may be an Auger effect. If so, the rate of recombination of electrons for example would be proportional to their density and to the density of holes with which they could combine, and proportional to the sum of electrons and holes which could act as the 'third body' to carry away the energy of recombination. Hence we may expect

$$1/\tau \propto p(An + Bp)$$

where A and B represent the relative effectiveness of either electron or hole as the 'third body'.

Now if for some reason electrons are much more effective in this role, i.e. $A \gg B$, then

$$1/\tau \propto np \propto n_i^2$$

Thus a variation as $\exp E/kT$ would be expected, even when the majority carrier concentration does not change greatly, as in de Carvalho's (1957) experiments.

178

OTHER ELEMENTS

No INTENSIVE optical work has been carried out on the other elements. However, information which has been published since the author's previous review (Moss, 1952a), is given in this chapter.

13.1 GRAY TIN

The most interesting development in this material is the brief report of photoconductivity by Becker (1955). Films ∼30 μ thick were used at 5°K, where the limit of photoconductivity corresponded to an energy gap of 0·075 ± 0·005 eV. This value compares well with the range of recent determinations from electrical measurements. These values are:

(a) Hall effect, 0·064 eV. Kendall (1954).
(b) Hall effect, 0·094 eV. Kohnke and Ewald (1956).
(c) Hall effect, 0·085 eV. Becker (1955).
(d) Magneto-resistance 0·082 eV. Ewald and Kohnke (1955).
(e) Magnetic susceptibility 0·08 eV, Busch and Mooser (1953); Busch (1958).
(f) Hall effect using single crystals grown in mercury, 0·08 eV (Ewald and Tufte, 1958).

Goland and Ewald (1956) conclude that the electron mass is approximately equal to, and the hole mass is half of, the free electron mass. Since Becker (1955) could detect no transmission through a film 30 μ thick at either 5°K or 78°K, it is concluded that the absorption coefficient exceeded 2000 cm^{-1} at 20 μ and 3000 cm^{-1} at 10 μ.

13.2 ANTIMONY

Photoemissive measurements by Taft and Apker (1954) have confirmed that antimony is a semi-conductor if deposited as films <300 Å thick. The Fermi level was found to be ∼0·1 eV above the full band.

From the temperature dependence of the electrical properties Cohen (1954) finds activation energies of 0·07 eV at low temperatures and 0·13 eV at higher temperatures—in substantial agreement with earlier results by the author (Moss, 1952a).

Mooser and Pearson (1957) point out that the semi-conducting properties shown by amorphous antimony are in accord with their concept of a semi-conducting band. The relation of the amorphous and crystalline phases is discussed by Richter *et al.* (1954).

13.3 IODINE

Franks (1957) has observed that iodine is luminescent in the infra-red, the peak emission occurring at 1·45 μ—i.e. just beyond the absorption edge. The peak of the excitation spectrum was in the highly absorbing region at 0·68 μ, coinciding approximately with the dip observed in many of the photoconductive spectral response curves (Moss, 1952*a*).

CHAPTER 14

LEAD SULPHIDE, SELENIDE AND TELLURIDE

14.1 GENERAL PROPERTIES

THE lead chalcogenides have an important technological interest as highly sensitive infra-red detectors. As a result they have been the subject of much fundamental research in the form of both single crystals and layers.

Although PbO is a photoconductor (see Moss, 1952a) it is not of great interest, and will not be considered further at present.

The other three materials crystallize in the face centred cubic lattice, the lattice constants being:

$$PbS \quad 5 \cdot 97 \text{ Å}; \quad PbSe \quad 6 \cdot 14 \text{ Å}; \quad PbTe \quad 6 \cdot 34 \text{ Å}$$

They may be made synthetically by techniques such as those used by Lawson (1951, 1952), although this worker was not able to produce PbS of purity as high as that of the best natural galena. More recently, by growing crystals in a controlled vapour pressure of sulphur, Scanlon et al. (1956) have produced specimens with resistivities up to 1 Ω cm, i.e. approaching the intrinsic value at room temperature. Similar techniques have been used by Bloem (1956a) and phase diagrams of the lead–sulphur system have been given by Bloem and Kroger (1956).

Considerable difficulty has been experienced in interpreting the electrical properties of these materials, and the onus of establishing the true values of the activation energies has fallen largely on optical and photoelectrical measurements. For PbSe and PbTe, one can only say even now that the electrical measurements are not inconsistent with the optical ones at high temperatures, (see Smith, 1953, 1954; Chasmar, 1956; Shogenji and Uchiyama, 1957). For PbS, the work of Scanlon, (1953) and his co-workers (Scanlon and Brebrick, 1954; Scanlon et al., 1956) on their excellent synthetic crystals has done much to clear the picture. From a plot of the product of electron and hole densities (i.e. $np \equiv n_i{}^2$) against temperature, these workers find a zero temperature activation energy of 0·34 eV. From measurements on a p–n junction the author (Moss, 1955c) has obtained a value of 0·40 eV for the room temperature activation energy. This would be reduced slightly if the lower values of effective mass estimated in section 14.2.2 were used in the calculations.

A theoretical calculation of the band structure has been carried through by Bell *et al.* (1953). Though of necessity drastic simplifications had to be made to keep the problem within reasonable proportions, the results obtained seem realistic. *Figure 14.1* shows the results for the (1, 1, 0) direction, this being the direction which showed the smallest energy gap.

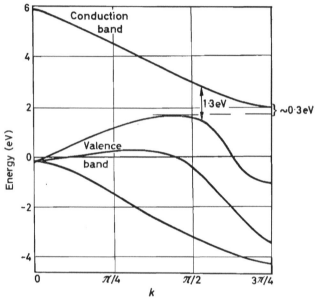

Figure 14.1. Theoretical E–k curves for PbS *in* (1,1,0) *direction*
(*Bell et al.,* 1953; courtesy *Royal Society*)

The curves show that the smallest separation between the conduction band minimum and valence band maximum is $\sim$0·3 eV, but the smallest vertical transition is considerably larger, namely $\sim$1·3 eV. A rough calculation made with a smaller than normal lattice constant showed that the energy gap should decrease on compression. Bell *et al.* (1953) also estimated the effective mass of a hole to be rather less than half the mass of a free electron.

The electrical properties of solid solutions of PbSe–PbTe have been studied by Kolomoets *et al.* (1957).

14.2 ABSORPTION

The first transmission measurements on crystals of these materials were made by Paul *et al.* (1951) on natural galena of high purity. A detailed examination of the transmission of synthetic crystals of all three materials was subsequently carried out by Gibson (1952*a*). Transmission measurements on PbS have also been reported by Clark and Cashman (1952) and Paul and Jones (1953).

14.2.1 Absorption edge region

The results of Gibson (1952*a*) for the three materials are shown in *Figures 14.2, 14.3* and *14.4*. No difference was observed between

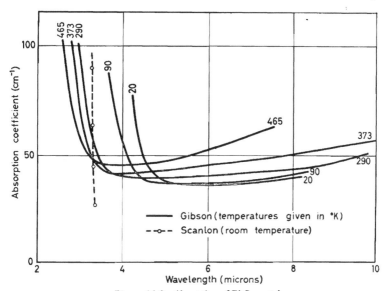

Figure 14.2. Absorption of PbS crystal

spectra for *n* and *p*-type specimens in the region of the absorption edge, although at long wavelengths it appears that *p*-type material always absorbs more strongly. Gibson's values for the position of the room temperature absorption edges—taken arbitrarily at an absorption coefficient ~100 cm^{-1} are:

PbS 0·39 eV;   PbTe 0·32 eV;   PbSe 0·26 eV

The actual absorption edges should no doubt be steeper than shown in *Figures 14.2, 14.3* and *14.4*, as Gibson's measurements

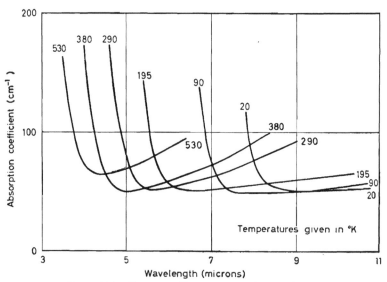

Figure 14.3. Absorption of PbSe *crystal.* (*Gibson, 1952a;*
courtesy *Physical Society, London*)

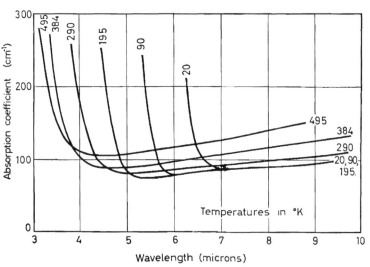

Figure 14.4. Absorption of PbTe *crystal.* (*Gibson, 1952a;*
courtesy *Physical Society, London*)

were made using very low resolution ($\sim$0·3 μ bandwidth). Measurements on PbS with better resolution and much thinner specimens have been made by Scanlon (1958*a*), some of whose results are shown in *Figure 14.2*.

Scanlon's results cover the whole of the absorption edge, and extend to levels well up on the absorption plateau shown in *Figure 14.5*. In the 1–2·5 μ region they show the same spectral dependence as *Figure 14.5*, but are about a factor of 2:1 lower. From these results the photon energy at maximum slope of the absorption edge can be found. It is 0·42 eV. Above $K \sim 3000$ cm^{-1} it is found that K^2 increases linearly with $h\nu$. Scanlon (1958*a*) interprets this as showing that the absorption is due to direct transitions, with an energy gap of 0·41 eV. Similar analysis (Scanlon, 1958*b*) gives 'direct' gaps of 0·29 and 0·32 eV for PbSe and PbTe.

For all three materials the absorption edge moves to longer wavelengths on cooling—contrary to almost all other semi-conductors. If the position of the edge in eV is plotted, the variation is found to be linear with temperature up to about 450°K, and for each material the shift has the same value of $+4 \times 10^{-4}$ eV/°C. This figure is virtually independent of the absorption level at which it is measured, and both the position of the edges and the shift are independent of impurity concentration.

The minimum absorption coefficient at 8 μ for PbS recorded by Gibson was 30 cm^{-1} for a specimen with 3×10^{17} cm^{-1} electrons. Paul *et al.* (1951) gave a value of 11 cm^{-1} at this wavelength for their very pure galena, with minimum absorption of 6 cm^{-1} at 4 μ.

Early work on the optical properties of films (Gibson, 1950) did not show the absorption edges which are present in the single crystals, but more recent work by Vernier (1953), Braithwaite (1955), Lasser and Levinstein (1954) and Levinstein (1956) has shown that in suitably prepared layers the absorption edges are well defined—in fact such layers make effective infra-red filters (Smith *et al.*, 1957).

14.2.2 *Free carrier absorption*

At long wavelengths *Figures 14.2, 14.3* and *14.4* show the rising absorption due to free carriers. For cooled specimens most of the absorption is frequency independent, but at elevated temperatures plots of K against λ^2 are approximately straight lines with intercepts on the absorption axis due to the residual (wavelength independent) absorption.

A general correlation between the absolute value of the absorption coefficient at long wavelengths and the carrier concentration was observed by Gibson (1952a), but as the results were scattered he preferred to analyse them in terms of the variation of absorption with variation in measured mobility at various temperatures. Now according to equation (2.33)

$$K\lambda^{-2} = Ne^3/4\pi^2c^3n\varepsilon_0m^2\mu$$

The data of *Figures 14.2, 14.3* and *14.4* show that $K\lambda^{-2}$ is approximately proportional to $1/\mu$ at the higher temperatures. With allowance for residual absorption, the experimental value of the parameter $nK\mu/N\lambda^2$ is:

PbS 19.3×10^{-12}; PbSe 18.6×10^{-12};

PbTe $12.5 \times 10^{-12}mks$ units

where the values of refractive index used are those given in section 14.3.2.

Theoretically for a free electron, equation (2.33) gives

$$nK\mu/N\lambda^2 = 0.53 \times 10^{-12}$$

Assuming that the whole of the discrepancy between experimental and theoretical values is due to the effective masses differing from the free electron mass, we obtain the following values for the effective masses:—

PbS 0.17; PbSe 0.17; PbTe 0.21 electron masses

As discussed in section 2.5 the theoretical expression for absorption is increased somewhat if the scattering mechanism is such that the collision time is energy dependent. If ionized impurity scattering is predominant—as it could well be for these materials as they were all far from intrinsic—the absorption would be increased by a factor $\gamma = 3.4$ and the above values of effective mass would be increased by $\gamma^{1/2}$ to become:

PbS 0.31; PbSe 0.31; PbTe 0.39 electron masses.

The values of effective masses determined from electrical data (see Moss, 1955b; Bloem, 1956a; Putley, 1955; Petritz and Scanlon, 1955) agree quite well with these latter estimates; although for PbTe the electrical data indicate

$$m^* = 0.2m \text{ to } 0.25m$$

which is nearer the value determined on the basis of lattice scattering.

Gibson (1952a) noted that the absorption of *p*-type material was in general several times that of *n*-type for the same impurity

186

concentration. This is partly explained by the fact that $\mu_e > \mu_h$ for all three materials. Also, from the analysis of Petritz and Scanlon (1955), it seems quite possible that $m_h < m_e$.

14.2.3 Theory of spectral shift

The measured value of the temperature dependence of the absorption edge is $+4 \times 10^{-4}$ eV/°C. As discussed in section 3.3 this is made up of two effects, a lattice expansion effect and a broadening of the energy levels by lattice vibrations. As the latter effect necessarily gives a negative shift, the dilatation effect alone must give a shift $>4 \times 10^{-4}$ eV/°C. There are indications that the broadening effect is relatively small, making the dilatation shift perhaps 5×10^{-4} eV/°C.

Putting this value and the measured expansion coefficients into equation (3.20) we obtain

$$\pm C_e \pm C_h = 12 \text{ eV}$$

As C_e and C_h are probably of comparable magnitude it is very likely that both positive signs should be used, giving

$$C_e \sim C_h \sim 6 \text{ eV}$$

We may use equation (3.19) to get an estimate of effective mass from these values of C—although as the equation should only be applied to valence crystals the results will be rather approximate. Estimating that

$$c_{11} = 0 \cdot 2 \times 10^{12} \text{ dynes/cm}^2,$$

as for selenium (see section 11.2.2), we obtain

$$C^2 \mu (m^*/m)^{5/2} = 2600$$

Putting $C_e = 6$ eV, and using the room temperature electron mobilities (Smith, 1954) namely PbS 600, PbSe 900, PbTe 1700 cm²/V.sec, we obtain for the effective masses

PbS $\sim 0 \cdot 4$;   PbSe $\sim 0 \cdot 4$;   PbTe $\sim 0 \cdot 3$ electron masses

These values compare quite well with those deduced in section 14.2.2.

14.3 REFLECTION MEASUREMENTS

Avery (1953, 1954) has determined the real and imaginary parts of the dielectric constant over the range $0 \cdot 5$–6 μ by analysis of the reflection of polarized light. The method used (Avery, 1952) was to measure the ratio of the reflection coefficients for incident

187

light polarized in, and perpendicular to, the plane of incidence. For each wavelength three angles of incidence were used, their values being chosen to bracket the pseudo-Brewster angle of minimum reflectance. The data were analysed graphically making use of sets of curves of n and k computed for the angles of incidence to be used.

14.3.1 Absorption

The results obtained for the three materials are shown in *Figures 14.5, 14.6* and *14.7*. It will be seen that in all cases the

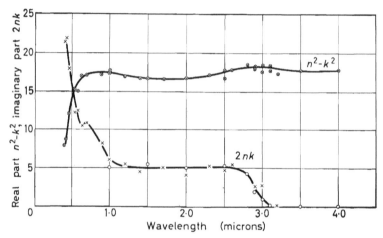

Figure 14.5. Real and imaginary parts of the dielectric constant of PbS. *(after Avery—unpublished data)*

absorption is characterized by a peak in the visible or ultra-violet regions, plus a plateau of virtually constant level extending over a waveband of several microns and terminating quite steeply in the neighbourhood of the absorption edge. It was established that for PbS the height of this plateau region did not vary significantly with impurity contents ranging from 6×10^{16} to 6×10^{19} cm^{-3}, so that the absorption tail is clearly characteristic of the PbS itself.

In view of the calculated band structure shown in *Figure 14.1* it seems reasonable to attribute the tail band to indirect transitions across the gap, and the rise in absorption below 1 μ with the onset of vertical transitions. It has been noted by Smith (1954) that at lower absorption levels (i.e. below 200 cm^{-1}) the cube root of the absorption varies linearly with frequency. Thus equation

188

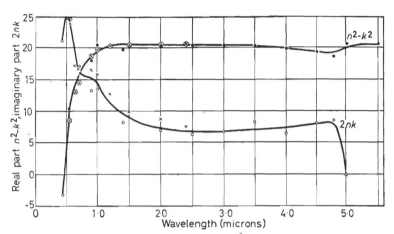

Figure 14.6. Real and imaginary parts of the dielectric constant of PbSe.
(after Avery—unpublished data)

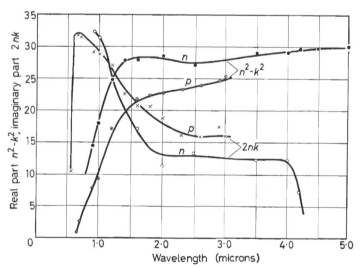

Figure 14.7. Real and imaginary parts of the dielectric constant of PbTe.
(after Avery—unpublished data)

(3.10) applies and theory indicates indirect transitions. Scanlon (1958) however, considers that direct transitions start at 0·41 eV.

Figure 14.7 shows that there is a significant difference between results for n and p-type samples of PbTe. No detailed investigation of this effect was made, but it may well be related to the known difference in surface properties of p and n material (Gibson, 1952b).

Reasonable agreement is obtained between values of K calculated from *Figures 14.5, 14.6* and *14.7*, and those determined from the transmission of thin films by Gibson (1950) throughout the region of very high absorption. In the plateau region however, the reflection data show K values about three times higher than those determined from transmission of films by Lasser and Levinstein (1954), and Levinstein (1956) and transmission of thin single crystals by Gibson (1952a), and Scanlon (1958a). In the 1·5–4 μ region the latter data are preferred to the reflection results.

At an absorption level of half the plateau value, Avery (1954) found for PbS that the edge moved by 4×10^{-4} eV/°C for temperatures between 20 and 150°C, the shift becoming larger at higher temperatures as the tail band steadily disappeared.

A rough representation of the optical behaviour may be made by fitting two or three classical absorption bands (see equations 2.6 and 2.6a) to the measured curves. For PbTe a band of oscillator strength 4 per molecule at 0·68 μ and another of strength 0·1 at 2 μ give an approximate fit. For PbSe, it is preferable to use three bands, namely 0·48 μ, $f = 3$; 0·85 μ, $f = 0·22$ and 3 μ, $f = 0·04$. For PbS, two bands of strengths 8 and 0·02 per molecule at 0·3 μ and 2 μ give a reasonable fit, although as the peak of the $2nk$ curve is not included in the measurements the estimation of the strength of the short wavelength band cannot be very good. This assessment is sufficient to show that in all cases the tail band oscillator strength is only 1 or 2 per cent of that of the main band, which again supports the view that the tail bands arise from indirect transitions. A total oscillator strength of about four for the range of spectrum measured would seem reasonable considering what is known of the electronic structure (Bell *et al.*, 1953).

Yoshinaga (1955) has measured the normal reflectance of PbS and PbSe at very long wavelengths. Both materials show a reflection minimum (of 11 per cent reflectivity) at about 45 μ, immediately before a rapid rise to high values of reflectivity which persist for wavelengths up to at least 200 μ. These effects are presumably caused by free carriers as discussed in section 2.5. From equation (2.36a) about 2×10^{18} cm^{-3} carriers would be required.

190

14.3.2 Refractive index

Figures 14.5, 14.6 and *14.7* show that there is little change in $n^2 - k^2$ beyond 1 μ (1·5 μ for PbTe). Since in the tail band k is only about unity, the change in n throughout the tail band—and on crossing the absorption edge—is small. The average values of refractive index for several specimens of each material are given by Avery (1953) in *Table 14.1*.

Table 14.1—*Refractive Indices at* 3 μ

PbS	PbSe	PbTe
4·10 ± 0·06	4·59 ± 0·1	5·35 ± 0·1

These values are, of course, very large. With the exception of tellurium the value for PbTe is the highest index measured in a non-absorbing region.

It should be noted that the refractive index increases in order of increasing molecular weight—in contrast to the absorption edge wavelength which is greatest for PbSe. The relation $n^4/\lambda_{1/2} \sim 77$, given in section 3.7.1 applies fairly well for PbS and PbSe—PbTe is the misfit.

Avery (1954) determined the temperature dependence of the refractive index of PbS over the temperature range 20–300°C. He found that the index decreased steadily on heating at a rate of 6×10^{-4} per °C. From the relation given by equation (3.25) namely

$$n^4 E_g = \text{constant}$$

we would expect the temperature dependence of the absorption edge and refractive index to be related by,

$$\mathrm{d}n/\mathrm{d}T = -(n/4E_g)\,\mathrm{d}E_g/\mathrm{d}T$$

From measured data for the edge shift we thus expect $\mathrm{d}n/\mathrm{d}T = -10^{-3}$ per °C, which is of the same sign and order of magnitude as the observed dependence.

Levinstein (1956) found that the refractive index of PbTe films could be changed considerably by oxidation, and also observed that the refractive index of PbTe increased on cooling.

14.4 INTERDEPENDENCE OF OPTICAL CONSTANTS

The data available for PbTe is sufficiently extensive to enable a check of the inter-relation of the optical constants to be made. In *Figure 14.6* the value of $2nk$ for the p-type material is small at the short wavelength limit of measurement and it can readily be extrapolated to zero. At long wavelengths, the edge determined on

191

the n-type sample may be used to complete the absorption curve of the p-type specimen.

From equation (2.29) the zero frequency refractive index is

$$n_0{}^2 - 1 = -\frac{2}{\pi} \int_0^\infty (2nk) \frac{\mathrm{d}\lambda}{\lambda}$$

Integrating this function graphically it is found that

$$(\pi/2)(n_0{}^2 - 1) = 46, \qquad \text{or} \qquad n_0 = 5\cdot5$$

This value is in good agreement with the observed long wavelength refractive index. It was shown in section 2.4.3 that the integral of the absorption coefficient also using the relation

$$n_0 - 1 = \frac{1}{2\pi^2} \int_0^\infty k \, \mathrm{d}\lambda$$

resulted in a refractive index value in good agreement with experiment.

On cooling, the absorption edge moves to longer wavelengths, thus increasing the area under the absorption curve. This should give an increased refractive index, as is observed experimentally.

14.5 PHOTO-EFFECTS

14.5.1 *Photoelectro-magnetic measurements*

The first observation of a photo-effect in bulk PbS was a P.E.M. measurement on Wisconsin galena of high purity (Moss, 1953a). From measurements of the absolute magnitude of the short circuit photocurrent for a known photon flux the quantum efficiency was determined using equation (4.55). The values obtained ranged from 40–150 per cent. In view of the inevitable inaccuracies in the experiments and the uncertainties in the values of mobility used in the calculations it seems probable that the quantum efficiency is always close to 100 per cent in this material.

A measurement of the spectral distribution of sensitivity (*Figure 4.4*) showed that for constant photon irradiation the sensitivity was sensibly unchanged between 1 and 2·8 μ, with $\lambda_{1/2} = 3\cdot02$ μ. The theory of section 4.4.7 shows that the $\lambda_{1/2}$ point corresponds to the condition $K(D\tau)^{1/2} = 1$. From the results of Gibson (1952a) we find $K = 150$ cm^{-1} at 3·02 μ. From average values of room temperature mobility in natural galena, $D \doteqdot 10$ cm^2/sec for either type of carrier. Hence, the above expression gives $\tau \doteqdot 5$ μsec.

By virtue of the slight dependence on lifetime of the effect, and by making use of the reversal of the effect with magnetic field, it was possible to make P.E.M. measurements on a variety of specimens. By selecting the best of these specimens it was possible to observe photoconductive effects, and photo-voltaic effects at metal probes. Lifetimes in the pure galena specimens were determined by the following methods: (*1*) Ratio of photoconductive to P.E.M. response, using equation (4.75), (*2*) diffusion length measurements by scanning light spot, and (*3*) observation of the rise time of photocurrent on illumination with the light of a spark gap. In addition the lifetimes of impure materials (down to $<10^{-9}$ sec) could be determined from the absolute magnitude of the P.E.M. effect, when none of the other techniques was usable.

The observed values (which range from 6×10^{-10} to 9×10^{-6} sec) are shown in *Figure 4.3*. The approximate correlation with the square of the resistivity and hence the square of carrier concentration suggests that a three body recombination process such as the Auger effect is determining the lifetime. This would be in agreement with Pincherle's (1955) theoretical estimates. More recently however, Scanlon (1957) using techniques described by Brebrick and Scanlon (1957), has observed some degree of correlation between lifetime and etch pit density and hence dislocations. The correlation is again not very specific and it is possible that both mechanisms contribute significantly, each becoming of major importance when the appropriate density (of impurities or dislocations) is large. It is noteworthy that with the highest dislocation density measured by Scanlon (1957), namely 10^{7} per cm^2, τ was reduced only to 10^{-7} second, whilst for very impure specimens Moss (1953*a*) found lifetimes one hundred times smaller than this. The longest lifetime found by Scanlon (1957) was 20 μsec for a specimen $\sim$0·3 Ω cm with 10^5 etch pits per cm^2.

The radiation lifetime of PbS calculated by Mackintosh (1956) using the theory of Chapter 6 is 40 μsecs—with a possible error of 5 : 1 either way. The more accurate value by Scanlon (1958*a*) is 63 μsec. For the purest material therefore, direct electron-hole recombination with photon emission must be. quite important. Recombination radiation from PbS in the region of the absorption edge has been observed by Galkin and Korolev (1953) and Garlick and Dumbleton (1954). The latter workers found that the emission band moved from the 2·6 μ region at room temperature to $\sim$3·3 μ at $-90°$K. For PbSe and PbTe Scanlon (1958*b*) calculates radiation lifetimes near 7 μsec.

14.5.2 Photo-voltaic effects

Photo-voltaic effects at metal probes have been studied for all three materials, while for the sulphide and selenide some observations have been made on *p–n* junctions.

193

Fischer *et al.* (1938) measured spectral sensitivity curves using Sardinian galena. For two specimens they obtained $\lambda_{1/2}$ values of 3·0 μ which are very close to the P.E.M. result. Lawrance (1951) recorded very high sensitivities with PbS photo-voltaic cells. For a *p–n* junction which was grown accidentally, Mitchell and Goldberg (1953) measured a spectral sensitivity curve peaking at 3 μ. Starkiewicz *et al.* (1957) found $\lambda_{1/2} = 2·9$ μ for a PbS *p–n* junction detector, with a response time <1 μsec.

The spectral sensitivity of a PbSe photodiode has been measured by Gibson (1952*a*) at room and liquid air temperatures. The threshold wavelengths were:

$$\text{Room temp.}—\lambda_{1/2} = 4·6 \ \mu; \qquad 90°\text{K}—\lambda_{1/2} = 6·9 \ \mu.$$

These results correspond to a spectral shift of $4·5 \times 10^{-4}$ eV/°C. From equation (4.31) the condition for the $\lambda_{1/2}$ point is

$$K(D\tau)^{1/2} = 1$$

The values of absorption coefficient given by Gibson (1952*a*) at the above wavelengths are 400 cm^{-1} and 90 cm^{-1} respectively. The mobility (of either electrons or holes) is near 1200 cm^2/V.sec at 290°K and 12,000 cm^2/V.sec at 90°K. Hence the estimated lifetimes are:

$$290°\text{K}—0·2 \ \mu\text{sec}; \qquad 90°\text{K}—1 \ \mu\text{sec}$$

These are the only available estimates of lifetime in bulk PbSe at present. They are fairly close to the calculated radiative lifetime of intrinsic material at room temperature of 0·3—1·2 μsec (Mackintosh, 1956). Natural *p–n* junctions have been detected in PbSe by Goldberg and Mitchell (1954).

A study of PbTe photodiodes has been carried out by Gibson (1952*b*). Spectral sensitivity curves measured at 90°K showed approximately constant sensitivity (per incident photon) for the wavelength range 1 μ–4 μ, with a slight peak at 5 μ immediately preceding the sharp fall in sensitivity in the absorption edge region. For a series of cells, using both pressure probes and soldered junctions, the following $\lambda_{1/2}$ values were observed:

$$4·9, \ 5·2, \ 4·8, \ 4·9, \ 5·3, \ 4·8, \ 5·2 \text{ and } 5·5 \text{ microns}$$

The average value is thus $\lambda_{1/2} = 5·08$ μ for 90°K. The spectral shift was about $4·5 \times 10^{-4}$ eV/°C.

At the maximum $\lambda_{1/2}$ value recorded the absorption data (*Figure 14.4*) show that $K = 150$ cm^{-1}. Using the relation $K(D\tau)^{1/2} = 1$, with D corresponding to a mobility of 25,000 cm^2/V.sec at 90°K, the carrier lifetime is $\tau = 0\cdot2$ μsec. This is the maximum value—the shorter wavelength cells would have considerably shorter lifetimes. Gibson (1952b) states that response times were <1 μsec. The radiation lifetime (see Chapter 6) has been calculated by Mackintosh (1956). He obtained $0\cdot8$ μsec., with a possible error of 3 : 1 either way for intrinsic material at room temperature.

14.5.3 *Photoconductivity in layers*

In their important role as sensitive infra-red detectors the lead chalcogenides are used in the form of thin layers, produced by vacuum evaporation, and in the case of PbS, by chemical deposition.

The preparation and sensitization of these layers is still more of an art than a science. The considerable literature on the subject has already been reviewed by the author (1955b, 1958a). Additional papers on this aspect have been published by Checinska and Sosnowski (1954), Pfeiffer (1956), Harada and Minden (1956) and by Munsch (1954). Usually the sensitive layers are made, processed and kept *in vacuo*, but in the case of chemical layers of PbS the detector is generally left open to the air or covered with a protective layer of varnish. PbS and PbSe layers can be made with considerable sensitivity at both room and liquid air temperatures, but it is only under exceptional circumstances that PbTe has significant sensitivity above 180°K.

Under the electron microscope the layers are seen to consist of close packed aggregates of oblong microcrystals of length $0\cdot1$ μ–1 μ, the size in general decreasing during the oxidizing sensitization processes (Moss, 1955b; Feltynowski *et al.*, 1954).

Spectral sensitivity curves for various temperatures are shown in *Figures 14.8, 14.9* and *14.10*. The $\lambda_{1/2}$ values in microns are given in *Table 14.2*.

Table 14.2—*Characteristic Wavelengths* $(\lambda_{1/2})$ *for Layers*

	295°K	90°K	20°K
PbS	2·9	3·8	4·1
PbTe	3·9	5·1	5·9
PbSe	5·0	7·1	8·2

The marked shift of the wavelength limit on cooling gives a significant increase in the useful range of the detectors. In accord with the absorption data, all three types of layer give a spectral shift of $+4 \times 10^{-4}$ eV/°C. For layers which are not suitably processed, the threshold wavelengths can be considerably less than those given in *Table 14.2* (Levinstein, 1956).

The spectral response of layers is also a function of thickness because the primary absorption process is thickness dependent.

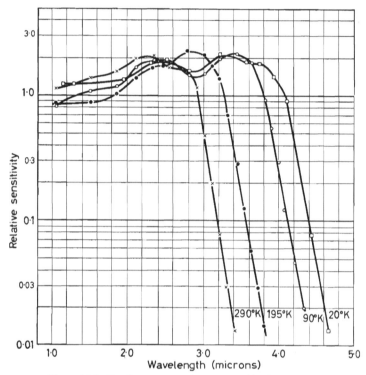

Figure 14.8. Equal energy spectral sensitivity of lead sulphide cell.
(Moss, 1949b; courtesy Physical Society, London)

In practice, if the layers are sufficiently thick to absorb nearly all the radiation throughout the plateau regions shown in *Figures 14.5, 14.6* and *14.7*, then the fact that the absorption edge is steep means that the threshold wavelength will be virtually thickness independent. For PbS and PbTe for example, layers of normal thickness ($\sim$1 μ) give spectral sensitivity curves which are substantially the same as those obtained with crystals a thousand times as thick.

For PbSe, however, some of the early photoconductive detectors were sensitive for wavelengths only up to $\sim$3·5 μ, probably because

196

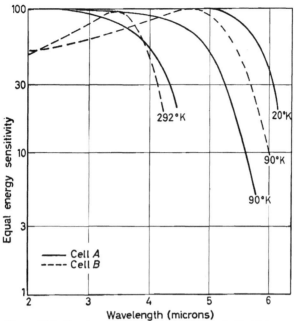

Figure 14.9. *Equal energy spectral sensitivity of lead telluride layers.*
(*Moss*, 1953d; courtesy *Research, London*)

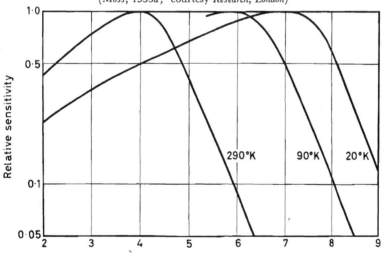

Figure 14.10. *Equal energy spectral sensitivity of lead selenide layer.*
(*Moss*, 1951; courtesy, *Physical Society, London*)

197

they were too thin to give adequate absorption. A detailed study of the influence of thickness on the spectral sensitivity of PbSe layers has been made by Humphrey and Petritz (1957a). This work and the transmission measurements of Braithwaite (1955) show that the high absorption coefficients indicated by the single crystal reflection measurements of *Figure 14.6* are not realized in PbSe films, the disparity being much greater than that for PbTe (see section 14.3). The estimated values of absorption coefficient in the edge region are shown in *Table 14.3*.

Table 14.3—PbSe *Absorption Coefficients* (cm^{-1})

	Type of Measurement	4·0 μ	4·5 μ
Avery	(crystal reflection)	25,000	20,000
Braithwaite	(film transmission)	3,000	800
Humphrey	(film photoconductivity)	1,000	200
Gibson	(thin crystal transmission)	1,800	700

There is reasonable agreement between the second and fourth lines in the table, particularly if Gibson's value at 4·0 μ is increased somewhat to allow for the fact that the measurement was made at a very high absorption level using low spectral resolution. On this basis, film thicknesses of 2 μ and 10 μ are required to give $\lambda_{1/2} = 4\cdot0$ μ and 4·5 μ respectively. Humphrey and Petritz (1957a) show that the shape of the absorption curve which they deduce from their photoconductive measurements is consistent with indirect transitions.

The effect of various gases and vapours on the spectral sensitivity of PbSe, and the theory of the sensitizing mechanism, have been studied by Jones (1957), Humphrey and Scanlon (1957) and by Humphrey and Petritz (1957b). Sensitization of cooled PbSe layers by strong illumination is reported by Schwetzoff (1957).

Time constants of modern sensitive PbS detectors are usually in the range 20–150 μsec for vacuum evaporated cells and several hundreds of microseconds for chemical layers. For the latter, the response time rises to $\sim10^{-2}$ second at liquid air temperatures, and these cells show a strong correlation between sensitivity and τ (Jones, 1953), as would be expected from simple photoconductivity theory. Lummis and Petritz (1957) found that in many sensitive layers the dependence of noise on frequency showed time constants of 100–300 μsec, in close agreement with the photoconductive response times.

For cooled PbTe cells, Levinstein (1956) quotes $\tau = 10^{-6} - 10^{-4}$

sec. For this material, no correlation between τ and sensitivity is observed and the most sensitive layers can have response times as short as 1 µsec.

Uncooled PbSe layers have fast time constants. Measurements by the author (1955b) give 0·5–1·5 µsec. Humphrey and Scanlon (1957) quote $\tau \sim 1$ µsec at room temperature and 15–30 µsec at $-195°$C.

14.5.4 Theories of photoconductivity in layers

Two theories of photoconductivity in these layers have been proposed. They are

(*1*) Single crystal recombination theory. Here the equilibrium excess carrier concentration on irradiation is determined by the respective rates of generation and recombination of hole-electron pairs, and the photoconductivity is taken to be directly proportional to the increase in the *number* of carriers, the mobility being sensibly the same as in the dark. The quantum efficiency will never exceed, but will be near to, unity.

(*2*) Barrier modulation theory. Here it is postulated that potential barriers are set up between microcrystals during layer manu- facture, and that these barriers are modified by the production of a few photoelectrons in their vicinity so that large numbers of carriers already present are permitted to flow across the layer. Thus the carrier *mobility* should increase markedly with only slight changes in overall carrier concentration. The apparent quantum efficiency could be much greater than unity.

The evidence is now conclusive in favour of the 'numbers' theory. Levy (1953) has shown by Hall effect measurements that for PbTe layers a change in numbers of 25:1 can be obtained with <2:1 change in mobility. Similar measurements by Woods (1957) on PbS have shown that of a given change of resistance on illumi- nation, <6 per cent is attributable to mobility variation. Smollett and Pratt (1955), Wood (1956) and Humphrey and Petritz (1957b), interpret the sensitization process in terms of single crystal theory, not the barrier model. Spenser (1958) concludes that the quantum efficiency is approximately unity.

There seems no theoretical necessity for the presence of potential barriers in order to obtain the sensitivities actually observed in the best cells. In fact barrier modulation, although it can increase the responsivity of a cell, may well result in worse sensitivity because the primary quantum efficiency is, by definition, small. Such barriers, however, may be necessary to explain the electrical

199

characteristics of oxidized PbS layers (Mahlman, 1956; Slater, 1956). Rittner (1956) also concludes that the barriers serve to reduce the mobility but are insensitive to light. Rittner and Fine (1955) have shown that the photo-effect is confined to the illuminated area, the drift of the carriers being $\ll 1$ mm.

The presence of actual potential barriers in layers has been demonstrated by observation of photo-voltaic effects on scanning with a sharply focussed light spot (Dutton, 1956). Such photo-voltaic effects are greatly enhanced if layers are subjected to a polarizing current during preparation. (Berlaga and Strakhov (1954) obtained an e.m.f. of 2 V from such a PbS photo-voltaic layer) or by use of trivalent metals for electrodes (Bloem, 1956b).

Recent theories of noise in layers have been given by Petritz (1956), Bess (1956) and Lummis and Petritz (1957).

CHAPTER 15

CHALCOGENIDES OF ZINC, CADMIUM AND MERCURY

ALL the diatomic compounds (Zn, Cd, Hg) (O, S, Se, Te) are interesting semi-conductors and have been studied to some extent. The general properties of the whole group will be reviewed together, and then the three most interesting materials ZnS, CdS and CdTe will be discussed in more detail.

Much of the research work on these compounds has been stimulated by their many technological uses as phosphors and photoconductors. In their commercial applications they are generally used in the form of multicrystalline layers, but much of the modern research work has been carried out on single crystals.

Useful reviews of the photoconductive properties have been given by Bube (1955a) and Garlick (1956).

15.1 PREPARATION AND GENERAL PROPERTIES

Most of the materials have very high melting points and the preparation of single crystals is very difficult. The most widely used methods are vapour phase reactions with H_2S, H_2Se and

Table 15.1—Nearest Neighbour Distances in Crystals in Å

	O*	S*	Se	Te
Zn	(1·94)	2·36 (2·35)	2·45	2·64
Cd	(2·35)	2·52 (2·52)	2·62	2·79
Hg		2·53	2·63	2·80

* Figures in brackets are not zinc blende structure

H_2Te (see Hamilton, 1958), or sublimation and recrystallization—often under considerable pressure. The 'heavy' compounds CdTe, HgSe and HgTe have lower melting points and crystals can be grown from the melt.

Most of the crystals have the zinc blende structure. The nearest neighbour distances are shown in *Table 15.1*. For comparison, the interatomic distance in Si is 2·35 Å.

In mixtures of ZnS and ZnSe the lattice constant increases linearly with the percentage of ZnSe molecules over the whole range, provided that the crystals are prepared at 850°C (Herman et al., 1957).

The binding forces in Wurtzite type crystals, including ZnO, have been discussed by Keffer and Portis (1957).

15.2 Electrical Properties

These compounds cover a wide range of electrical conductivities, the 'light' compounds being almost insulators if pure, while the 'heavy' compounds show semi-metallic properties. For some of the materials, the conductivity can be varied over a particularly wide range—for example CdS may be prepared with resistivity between 10 and 10^{12} Ω cm.

For some of the compounds reliable values of the activation energy have been obtained from measurements of the temperature dependence of conductivity or Hall constant. For ZnS Piper (1953) observed intrinsic conductivity in a single crystal, and with allowance for the probable temperature dependence of mobility deduced an energy value $E_0 = 3.67 \pm 0.1$ eV.

Appel and Lautz (1954) give the energy gap for CdTe as 1.43–1.57 eV, although some specimens show lower values, particularly at lower temperatures (Appel, 1954). Jenny and Bube (1954) quote 1.45 eV and have given the electron mobility as 300 cm²/V.sec and hole mobility as 30 cm²/V.sec for this material.

In HgSe and HgTe the energy gap is a very small fraction of 1 eV. In conjunction with their high mobilities ($\sim 10^4$ cm²/V.sec, Goodman and Douglas, 1954) this leads to large conductivities even in intrinsic material.

At temperatures near 1000°C ZnO shows a rapid exponential increase of conductivity (Weiss, 1952) which would indicate an activation energy of about 3.5 eV. Mollwo (1956) however, points out that the factors controlling this high temperature conductivity are not yet fully understood. The technique of using high resistance powdered samples (Ruppel et al., 1957) may enable better measurements of the intrinsic conductivity to be made. Hutson (1957/8) gives the electron mobility as 180 cm²/V.sec at 300°K, and the donor activation energy as 0.05 eV. The probable energy levels of this material have been discussed by Harrison (1954).

The electron mobility in both CdS and CdSe is ~ 100 cm²/V.sec (Bube, 1955a). The effective electron mass in CdO is $\sim 0.1m$ (Wright and Bastin, 1958).

202

15.3 OPTICAL PROPERTIES

With the exception of the three materials which will be discussed more fully later, most work has been done on the oxides of Zn and Cd.

According to Weiss (1952) the absorption coefficient of thin ZnO layers (prepared by oxidizing evaporated Zn) rises from negligible values at 3·2 eV to near 3×10^5 cm^{-1} at 3·3 eV, thereafter increasing slowly to 4×10^5 cm^{-1} at 5·5 eV. As these absorption coefficients are so large it seems probable that this is the main absorption edge, although, if so, it is surprising that it is at longer wavelengths than for ZnS. For ZnO crystals however, Mollwo (1956) shows spectra with the absorption edge at 3·5 eV. At this wavelength, although the absorption coefficient is rising rapidly, it has still reached only 500 cm^{-1}. An upper limit $\sim$6 eV for the activation energy has been suggested by Miller (1951).

ZnO shows birefringence, the refractive indices being 2·01 and 2·03. At wavelengths near 0·4 μ the dispersion is very large (Mollwo, 1956). From such low index values the relation $n^4/\lambda_c = 77$ would lead to an expected absorption edge near 0·2 μ, or an energy gap of 5–6 eV. Garlick (1956) has also suggested that the optical absorption edge should be at about 6 eV. The dielectric constant at a wavelength of 1 cm is 8·5 (Hutson, 1957), which is much larger than the (refractive index)^2, indicating strong lattice absorption in the infra-red region.

Measurements of absorption in CdO have been made by Stuke (1954). The edge is not particularly steep, the photon energies for absorption coefficients of 10^4 and 10^5 cm^{-1} being:

10^4 cm^{-1}, 2·19 eV and 2·35 eV at 100°C and −170°C

10^5 cm^{-1}, 2·64 eV and 2·72 eV at 100°C and −170°C

At an absorption level of 5×10^4 cm^{-1}, Dunstadter (1954) gives $2·5 \pm 0·1$ eV depending on the preparation temperature. Stuke (1954) finds a 'knee' on his spectral absorption curve at $2·8 \pm 0·2$ eV and he identifies this with the activation energy. From the above figures the shift of the edge is

$$-6 \times 10^{-4} \text{ eV/°C at } 10^4 \text{ cm}^{-1}$$

The absorption edges for mixtures of the Zn compounds have been measured by powder reflection techniques (see section 3.2.2) by Larach et al. (1957) who find that for ZnS–ZnSe mixtures the optical activation energy varies linearly with concentration, but

203

that selenide–telluride and sulphide–telluride mixtures show a minimum energy gap at ∼0·7 mole fraction of telluride.

Mercury selenide and telluride show ill-defined absorption edges at very long wavelengths. It is possible that these edges are dependent on impurity concentration (see section 16.6) as is observed for InSb. The high mobilities reported for these materials, indicating low effective masses, make this quite likely. The optical and electrical properties of HgTe have been described by Lawson et al. (1958).

The absorption edge of very thin specimens of CdSe single crystals has been investigated by Gross and Sobolev (1956) who find that at 4°K there are several well defined lines in the region 0·665 to 0·68 μ. These lines are attributed to excitons.

Bube (1955a) states that the absorption edge of ZnTe is at 2·15 eV. The refractive index of this material is 3·56 at 0·59 μ. As this measuring wavelength is right on the absorption edge it is clear that the refractive index will be much enhanced and the long wavelength, non-dispersive index, will be considerably less than 3·56. Judging by the dispersion curves for CdS and ZnS (*Figures 15.2, 15.6*) it is estimated to be ∼3·0.

The refractive index of ZnSe is 2·89, again at 0·59 μ—giving an estimated long wavelength value ∼2·7.

The refractive index values are summarized in *Table 15.2*.

Table 15.2—Refractive Indices

	O	S	Se	Te
Zn	2·0	2·28	∼2·7	∼3
Cd		2·32		$<\sqrt{11}$
Hg	$\begin{cases} 2·37 \\ 2·5 \\ 2·65 \end{cases}$	2·58, 2·82 or 2·76, 3·06		

Estimates of the dielectric constants of powdered samples of the sulphides, selenides and tellurides of all three metals have been made by Kyryashkina et al. (1957). The values quoted for the heavier compounds are surprisingly high and it is not certain that this technique of measurement is valid for such materials.

15.4 PHOTOCONDUCTIVITY

Photoconductivity in most of these materials is characterized by the fact that the apparent quantum efficiency can be much larger than unity—that is, many unit charges pass between the electrodes for each absorbed photon. The mechanism of this effect is that when an electron-hole pair is formed the hole is rapidly trapped while the electron passes freely through the crystal. As this electron leaves the crystal at the positive electrode another enters at the negative electrode, and so on until the electron finally recombines with the trapped hole. Miller (1951) has recorded 'gains' in the region of 10^3 for ZnO, and values up to 10^4 have been observed for CdS.

Rose (1955) points out that this type of behaviour is more likely than the simple unit quantum efficiency behaviour in high resistivity semi-conductors. It is most marked in the 'lighter' compounds covered in this chapter, but it is probable that the majority of oxide photoconductors show the effects to some degree. Rose (1955) also points out that the meaning of lifetime has to be considered rather carefully in such materials.

Using the simple phenomenological definition of electron lifetime, $\Delta n = q\tau_e$, which simply states that the equilibrium increase in carrier concentration equals the generation rate times the mean life, we can write the photocurrent as $\Delta I = eq\tau_e/t$ where t, the transit time between electrodes of separation X is given by $t = X/E\mu$.

The ratio τ_e/t which is the apparent quantum efficiency, is generally referred to as the 'gain' of the photoconductor, since it gives the number of charges passing through the electrodes per quantum absorbed.

It should be noted that whereas the photocurrent in an applied electric field is determined only by the electron lifetime—the hole lifetime being assumed very much shorter—a photoelectro-magnetic current will be determined by the shorter of the two lifetimes, as it is only while the two carriers are moving together that the primary diffusion current is produced. Use of the P.E.M. effect is thus potentially important for determining these minority carrier lifetimes. Typically for CdS $\tau_e \sim 10^{-3}$ and $\tau_h \sim 10^{-8}$ seconds.

As a further complication the lifetime τ_e does not in general equal the rise and decay times of the photocurrent because the electron may be trapped in shallow trapping levels and released again several times before ultimate annihilation by a hole. This total 'excitation life' may thus be much greater than the actual time spent as a free electron in the conduction band, and it is only this latter time which determines the equilibrium photocurrent and thus equals τ_e.

In general the photocurrent is proportional to the irradiation,

205

particularly at low illumination levels. At higher irradiation levels the photocurrent usually increases less rapidly than the intensity, but CdSe and ZnSe show the interesting phenomenon of 'super-linear' photoconductivity, i.e. the photocurrent in some circumstances increases *more* rapidly than the intensity (Bube, 1956; Bube and Lind, 1958).

In addition to change in conductance on illumination, both ZnO and ZnS have been found to show the analogous effect of change in dielectric constant on illumination (Roux, 1956; Borissov and Kanev, 1955). That such an effect must occur follows of course from the inter-relation of the real and imaginary parts of the dielectric constant (see section 2.4).

The spectral response curves of photoconductivity in these materials usually exhibit a peak in the neighbourhood of the absorption edge, characteristic of high surface recombination as discussed in section 4.2. When these peaks are sharp, their wavelength can be taken to give the activation energy, although such energy values will usually be slightly on the high side.

Characteristic wavelengths determined from photoconductive spectra are summarized in *Table 15.3*.

Table 15.3—Characteristic Photoconductive Wavelengths

	μ	*Reference*
ZnO	Peak 0·4	Mollwo and Stockman, 1948
ZnO	$\lambda_{1/2}$ 0·39	Weiss, 1952
HgO	0·55	Dechene, 1939
ZnS	Peak 0·337	Piper, 1953
CdS	Peak 0·50	Bube, 1953
CdS	$\lambda_{1/2}$ 0·515	Bube, 1953
HgS	Peak 0·63	Shneider, 1956
HgS	$\lambda_{1/2}$ 0·57	Bergman and Hansler, 1936
ZnSe	Peak 0·48	Bube, 1955b
CdSe	Peak 0·725	Bube, 1955b
CdSe	$\lambda_{1/2}$ 0·78	Treu, 1939
ZnTe	Peak 0·8	Braithwaite, 1951
CdTe	Peak 0·88	Bube, 1955b
HgTe	$\lambda_{1/2}$ 3·1	Braithwaite, 1951

Bube (1955b) has measured the shift of the photoconductive spectrum over the temperature range 100–400°K, using the position of the sharp peak to derive the following values of energy shift:

ZnSe $-7\cdot2 \times 10^{-4}$ eV/°C   CdSe $-4\cdot6 \times 10^{-4}$ eV/°C

CdS $-5\cdot2 \times 10^{-4}$ eV/°C   CdTe $-3\cdot6 \times 10^{-4}$ eV/°C

The high performance attainable by radiation detectors made

of polycrystalline layers of these materials is illustrated by the figures given for CdSe by Schwarz (1951).

15.5 Zinc sulphide

Much of the early experimental work on zinc sulphide was carried out on evaporated layers or sintered powders, but in the last few years increasing work has been carried out on single crystals. Though these are still difficult to make and are not large, it has been possible to carry out many optical and electrical measurements on them. The energy level system of ZnS has been discussed by Williams (1957), and Shakin and Birman (1958). The latter conclude that both extrema lie at $\vec{k}000$.

15.5.1 Absorption

Transmission determinations of the absorption index k have been made on thin films by Hall (1956) and Coogan (1957). Their results, which are shown in *Figure 15.1*, are in substantial agreement, and may therefore be taken as characteristic of film absorption.

Measurements on single crystals, however, by Piper (1953), which are also given in *Figure 15.1*, show very much less absorption. These latter results are supported by further single crystal measurements by Czyzak *et al.* (1954) who find for their best specimens an absorption coefficient of only 20 cm^{-1} at 0·344 μ, and 10 cm^{-1} at 0·355 μ. It must therefore be concluded that the macroscopic ZnS lattice shows negligible absorption for wavelengths down to about 0·35 μ. For four specimens Piper (1953) found that at an absorption level of 10 cm^{-1} the photon energy was 3·56–3·60 eV, and he therefore took the optical activation energy to be 3·58 eV.

From the shape of his absorption curve, Coogan (1957) decided that it could be split conveniently into a main edge plus a small absorption band, as shown by the dotted lines in *Figure 15.1*. It will be seen that the dotted edge thus obtained for the main band joins fairly well onto the single crystal data. It must be concluded therefore that this subsidiary band is not characteristic of ZnS itself, but is a consequence of the structure of films of this material. It is however surprisingly intense, having a peak absorption coefficient $K \sim 10^5$ cm^{-1}.

Although the single crystal absorption edge shown in *Figure 15.1* looks very steep, Piper (1953) has shown by more detailed measurements that log K falls linearly with photon energy (or with wavelength) over a range of more than 100:1, the rate of fall being, $\mathrm{d}(\log K)/\mathrm{d}(h\nu) = 12\text{–}16$ eV^{-1}. This corresponds quite closely to a change of e:1 in K per kT increase in photon energy.

With such behaviour it is impossible to define a specific absorption limit at which absorption ceases entirely, and it is preferable—as outlined in section 3.2.1—to use the criterion of maximum slope of the spectral dependence of absorption coefficient. For films,

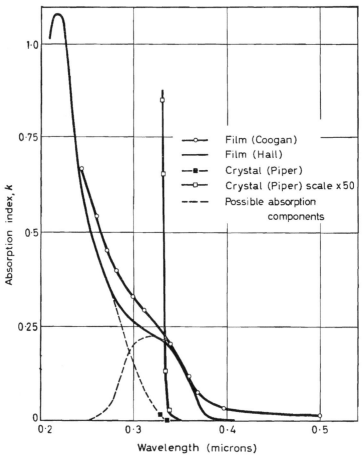

Figure 15.1. Absorption index for zinc sulphide

Figure 15.1 shows this wavelength to be 0·33 μ. For single crystals, Piper's results do not extend to sufficiently short wavelengths to identify this point accurately, but it will clearly lie near the shortest wavelength measured i.e. 0·33 μ or 3·75 eV. This is quite close to the thermal activation energy given in section 15.2, but in view

208

of the fact that the thermal activation energy is that appropriate to $0°K$—where the optical activation energy will be increased—there is possibly a significant difference between the two values.

Coogan (1957) measured the shift of the absorption edge at $k = 0.1$ to be -4×10^{-4} eV/°C between $-186°C$ and $20°C$. This value, however, applies to the additional (film) absorption band. Taking Coogan's results at $k = 0.4$, where the additional band should be unimportant, the shift becomes -7×10^{-4} eV/°C. Figures quoted by Moglich and Rompe (1942) correspond to approximately -5×10^{-4} eV/°C below room temperature, and -8×10^{-4} eV/°C above room temperature. Van Doorn's (1954) values are very similar.

At short wavelengths the absorption level is so high that the data obtained from films are probably valid for the single crystal lattice. *Figure 15.1* shows that maximum k occurs at 0.216 μ. From equation (2.19) the resonance frequency will correspond to a slightly longer wavelength than this, i.e. around 0.23 μ, giving $\omega_0 = 8.2 \times 10^{15}$. Treating the absorption as equivalent to a simple classical oscillator the expected long wavelength refractive index n_1 would be:

$$n_1{}^2 = 1 + Ne^2 f/m^* \varepsilon_0 \omega_0{}^2$$

Taking $\omega_0 = 8.2 \times 10^{15}$, $N = 5.2 \times 10^{28}$ atoms/m³, m^* equal to a free electron mass and $f = 1$ (i.e. equivalent to two per molecule) we obtain

$$n_1 = 1.85$$

This is about 25 per cent below the observed value, which is probably as good a measure of agreement as can be expected from such simple theory.

15.5.2 Refractive index

The results of several measurements of the refractive index of ZnS are plotted in *Figure 15.2*. The most reliable values are those obtained by Czyzak et al. (1957) on single crystals. These workers were able to grow crystals sufficiently large to make prisms although the resulting prism faces were only ~ 4 mm² in area. From their work the extrapolated long wavelength refractive index is given by

$$n_1{}^2 = 5.13 \qquad \text{or} \qquad n_1 = 2.27$$

The results of Hall and Ferguson (1955), obtained on films by measurement of fringes in reflection and transmission, show good agreement with the single crystal measurements. Coogan's (1957) results are also in reasonable agreement in the wavelength region where they overlap the single crystal work.

At shorter wavelengths there is a serious difference between the results of Coogan (1957) and Hall (1956). Those of Coogan (1957) are preferred since they show the type of behaviour expected from the absorption spectrum, on the basis of the inter-relation of the optical constants as discussed in section 2.4.3. These inter-relations show that a sharp absorption edge must result in a peak in the refractive index curve at approximately the wavelength of the edge.

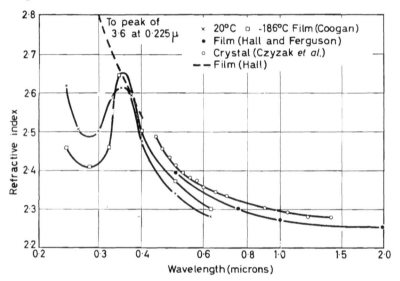

Figure 15.2. The refractive index of zinc sulphide

Also the peak in n should become sharper as the absorption edge steepens—as it would be expected to do on cooling. The results of Coogan (1957) show just such behaviour.

Similar analysis would indicate that the main peak of refractive index should occur near 0·225 μ—i.e. the wavelength of maximum slope in *Figure 15.1*. Coogan's measurements do not extend to quite such short wavelengths, but they indicate the proximity of a peak. The dispersion curve given by Hall (1956) does in fact show a peak at just this wavelength, so that his results are quali-tatively correct in this region. His value of the peak index of 3·6 however, seems very high.

Ramachandran (1947*b*) has shown that the refractive index increases with temperature. For temperatures of 100°C and above the change is $dn/dT = 0.84 \times 10^{-4}$ per °C at 0·62 μ, with somewhat

210

higher values at shorter wavelengths. On the basis of equation (3.25), this refractive index change would indicate a spectral shift given by

$$dE/E \, dT = -4 \, dn/n \, dT, \quad \text{or} \quad dE/dT = -5 \cdot 3 \times 10^{-4} \, \text{eV}/°\text{C}$$

This is the same sign and approximately the same magnitude as the observed change in the optical activation energy.

15.5.3 Reflectivity at long wavelengths

Yoshinaga (1955) has measured the reflectivity of a ZnS crystal for wavelengths beyond 20 μ. The results show a sharp Restrahlen band with a peak reflectivity of 86 per cent at 25·5 μ, preceded by a reflection minimum at 23·5 μ where the reflectivity is only $\sim$2 per cent. Beyond 45 μ the reflectivity is constant at 23 per cent indicating a long wavelength refractive index $n_0 = 2·84$ or $\varepsilon = 8·1$. Sinton and Davis (1954) find that this region of constant reflectivity extends to 240 μ. The above value of refractive index contrasts with the non-dispersive index in the near i.r. of $n_1 = 2·27$ or $n_1^2 = 5·13$.

The expected contribution of the Restrahlen band can be calculated from the equation (Burstein and Egli, 1955),

$$n_1^2 - \varepsilon = (Ne^2/9m\varepsilon_0\omega_0^2)(n_1^2 + 2)$$

provided m is taken as the reduced mass of the atoms, i.e.

$$1/m = 1/M_1 + 1/M_2$$

since it is vibrating ions which are causing the effect, and N is taken as the density of ion pairs. With these provisos we obtain

$$e = 0·9 \text{ free electron charges}$$

from the measured difference in dielectric constants quoted above.

This band may be analysed using the formulae for section 2.3. From equation (2.21) we find $\rho = 17$. Attempting to find the relative bandwidth we obtain values from $g = 9 \times 10^{11}$ radians/sec to $g = 25 \times 10^{11}$ radians/sec, depending on the reflection level used for the evaluation, indicating that the band is not a simple classical absorption band in shape. The resonance frequency is very close to $\omega_0 = 7·3 \times 10^{13}$ radians/sec, whichever value of g is used.

As the band has been shown to be more complex than that of a simple oscillator, any detailed evaluation would require the use of

211

the integral relation between the amplitude and phase of the complex reflectivity as outlined in section 2.4. This has been evaluated at one wavelength, that of peak reflectance, where it is found that $\sigma = 0.93$ and $\theta = 46°$ giving $k = 2.3$ and $n = 0.25$. These results are consistent with rough estimates which can be made using the condition for maximum reflectivity given by equation (2.21).

15.5.4 Faraday effect

Many years ago the Faraday effect in ZnS was determined by Becquerel (1877), the value quoted* being 0.225 minutes rotation/cm/G at 0.589 μ, or 65 radians/m per weber/m². Now from equation (5.10),

$$\theta = \frac{\omega \omega_c}{2c} \frac{dn}{d\omega} = \frac{Be}{2m^*c} \frac{\lambda}{d\lambda} \frac{dn}{d\lambda} \text{ radians/m}$$

Figure 15.2 gives $dn/d\lambda = 0.38$ at 0.589 μ, so that all parameters in the equation are known except the effective mass m^*, which may therefore be evaluated quite accurately. We find,

$$m^* = 1.1m$$

15.5.5 Photoconductivity

Measurements of photoconductivity in single crystals of ZnS have been made by Piper (1953). The spectral distribution of sensitivity for a crystal of about 1 mm² cross section is shown in Figure 15.3. It will be seen that the yield rises to 10 electrons per absorbed photon at 3.68 eV, or 0.337 μ.

The curve shows a very sharp peak near the absorption edge, characteristic of materials with very high surface recombination velocities. As shown by equation (4.39), the peak response in such circumstances should occur when $KL \sim 1$, i.e. when the absorption coefficient is about the reciprocal of the diffusion length. From Piper's absorption measurements on similar specimens, the absorption constant at 0.337 μ is $K = 500$ cm⁻¹.

In view of its crystal structure, ZnS must be expected to be birefringent (as for CdS) and so there will be two slightly differing absorption edges for directions along, or perpendicular to, the c axis. It is not certain that the absorption and photoconductive results of Piper (1953) are for the same crystal orientation, so that a specific calculation of diffusion length from the absorption constant at the wavelength of peak photoconductivity would not be justified.

Very marked anisotropy in photocurrents and electroluminescence

* See Handbook of Chemistry and Physics, for example.

has been observed by Lempici *et al.* (1957), differences up to 10^3 or 10^4 being observed for parallel and perpendicular directions. These effects are attributed to anisotropic stacking faults producing barriers within the crystals. In crystals with alternate regions of

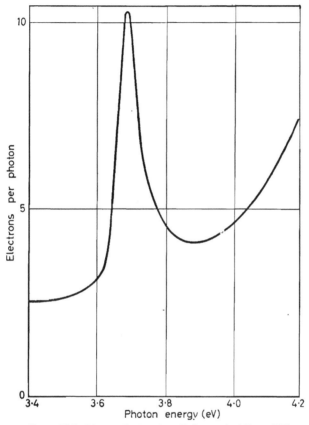

Figure 15.3. Photoconductivity in a ZnS *crystal.* (*Piper*, 1953; courtesy *American Physical Society*)

cubic and hexagonal ZnS, Merz (1958) has observed photo-voltages as great as 500 V.

Recent work on photoelectroluminescence in ZnS has been described by Kazan and Nicoll (1957), who have shown that effective solid state light amplifiers can be made using this effect. The correlation between luminescence and photoconductivity and their temperature dependence have been studied by Broser and

Broser-Warminsky (1955), and electroluminescence and thermoluminescence in single crystals have been discussed by Neumark (1956), and Piper and Williams (1958).

15.6 CADMIUM SULPHIDE

Much of the modern interest in cadmium sulphide stems from the work of Frerichs (1947), who grew crystals of CdS by the reaction

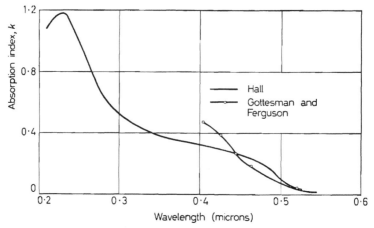

Figure 15.4. Absorption of cadmium sulphide films

of cadmium vapour with H_2S. He found that these crystals had a steep absorption edge at 0·51 μ, and a sharp peak in the photoconductive spectrum at wavelengths just beyond this where the effective quantum efficiency could reach $\sim 10^4$. The properties of CdS have been reviewed recently by Lambe and Klick (1958).

15.6.1 Absorption

Measurements of the absorption index made by Hall (1956), using transmission and reflection fringes from CdS films, are shown in *Figure 15.4*. The index peaks at 0·23 μ. This wavelength is not much greater than that for ZnS, so that the refractive index would be expected to be not much greater than for ZnS. This is confirmed experimentally, as discussed in the next section.

The long wavelength absorption edge lies close to 0·5 μ (i.e. the wavelength of maximum slope) but, as expected, it is not particularly steep in the film measurements. Results for single crystals however show that the absorption coefficient is rising steeply near 0·53 μ (Reynolds *et al.*, 1955). *Figure 15.5* shows transmission measurements on single crystals in the absorption edge

214

region at various temperatures. These results, which were obtained on plates ~0·1 mm thick, show that the edge is quite steep at room temperature and below. At the half maximum transmission points (where the absorption constant is ~100 cm⁻¹) the wavelengths and corresponding spectral shifts are as shown in *Table 15.4*.

Table 15.4

Temperature (°C)	Wavelength (μ)	Shift (eV/°C)
−196	0·492 ⎫	−5·5 × 10⁻⁴
20–27	0·517 ⎬	
110	0·534 ⎭	−8·4 × 10⁻⁴

Frerich's results show that the edge moves linearly with temperature from 20°C–500°C at the rate of −6·5 × 10⁻⁴ eV/°C. This latter value is preferred to the higher value tabulated above. Between 77°K and 4°K the shift is quite small (Klick, 1953). Such temperature dependence of the shift has now been observed in several materials.

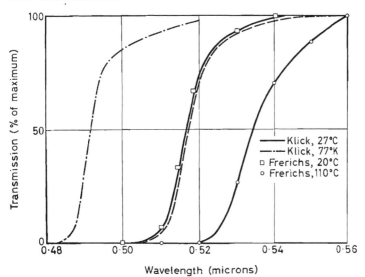

Figure 15.5. Transmission of cadmium sulphide crystals

Gobrecht and Bartschat (1953) have measured the transmission with polarized light. With the E vector parallel to the c axis the absorption edge lies at shorter wavelengths than when perpendicular to the c axis. The wavelength separation between these two

215

absorption edges is about 35 Å at both room and liquid air temperatures. More recently Bartschat and Gobrecht (1958) have studied the optical anisotropy at both low and elevated temperatures. D. L. Dexter (1958) finds reflectivity peaks at 4840 Å and 4870 Å and states that both absorption edges are exponential with slopes near $1/kT$.

At low temperatures Broude et al. (1957) have resolved a number of discrete lines in the spectrum of CdS crystals. These lines are more numerous for observations perpendicular to the c axis. Gross and Yakobson (1955) have observed as many as twenty lines near the absorption edge. Most of the lines lie between 4875 Å and 4856 Å, and have a separation of about 10 cm^{-1}. Gross and Yakobson (1956) have observed emission lines in the same spectral region. Emission lines have also been measured by Klick (1953), who finds that the lines are more clearly defined at 4°K and that they have a separation of 0·04 eV—i.e. many times the separation of the lines reported by the Russian workers.

Halperin and Garlick (1955) have observed additional absorption in the near i.r. when CdS crystals are excited by visible radiation.

Höhler (1949) has found that the absorption edge moves to shorter wavelengths with increasing pressure, the shift being approximately $1·7 \times 10^{-2}$ Å per atmosphere. He shows that this is equivalent to a temperature shift of $\sim 1 \times 10^{-4}$ eV/°C., so that only about a 1/5 of the temperature shift is attributable to the dilatation effect, and the major contribution must be due to the broadening term.

The theoretical expressions for spectral shift given in section 3.3 will not be strictly applicable to a compound like CdS, and in any case the relevant parameters are not known for this material. However, it is worthy of note that the linear expansion coefficient is rather small (6×10^{-6}—Höhler, 1949) which will tend to result in a small dilatation effect. Also this effect will be relatively small if the bottom of the conduction band and top of the valence band move in the same direction on expansion or compression.

The presence of considerable free carrier absorption in the 7–14 μ region is indicated by measurements by Hall and Ferguson (1955). Over this range the absorption coefficient is proportional to (wavelength)^2, the magnitude being $K = 0·2$ cm^{-1} at 10 μ. That such an absorption level is consistent with free carrier absorption may be seen from equation (2.32), namely $cnK = Ne^3/m^{*2}\varepsilon_0\mu\omega^2$. Putting $\mu = 100$ cm^2/V.sec and $m^* = m$ gives $N = 10^{17}$ cm^{-3}. Alternatively, if $\mu = 30$ cm^2/V.sec and $m^{*2} = 0·1m^2$, then $N = 3 \times 10^{15}$ cm^{-3}. These estimates of m^* and μ and the values of N

are consistent with actual observations on CdS crystals by Kroger *et al.* (1954) and cyclotron resonance work by R. N. Dexter (1958).

Klick (1953) states that there is a weak absorption band at 34 μ (with $K_{max} \sim 100$ cm⁻¹) and that the Restrahlen band should lie in the 40–45 μ range.

15.6.2 Refractive index

Czyzak *et al.* (1957) have carried out accurate measurements of refractive index on small prisms made from single crystals of CdS. Their results are shown in *Figure 15.6*, where it will be seen that

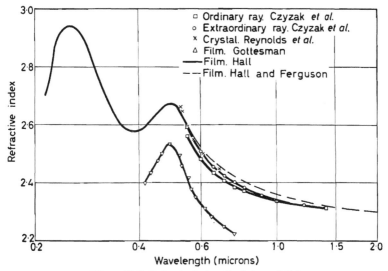

Figure 15.6. Refractive index of cadmium sulphide

there is slight birefringence, the index being higher when the light is polarized parallel to the *c* axis. Some measurements to rather shorter wavelengths by Reynolds *et al.* (1955) for light polarized parallel to the *c* axis are also shown. These results indicate a long wavelength refractive index $n_0 = 2.3$.

Measurements on films of transmission and reflection fringes by Hall and Ferguson (1955) are in quite good agreement with the single crystal results. Hall's (1956) film values are also in good agreement with the crystal data where they overlap. At shorter wavelengths the refractive index shows a peak at 0·5 μ, i.e. near the absorption edge, and a larger peak at 0·25 μ. Reference to *Figure 15.4* shows that these peak wavelengths are near those of

217

maximum slope of the absorption curve so that, as discussed for ZnS, the behaviour is in accord with deductions based on the inter-relation of the optical constants.

Results by Gottesman and Ferguson (1954) also show a re-fractive index peak at 0.5 μ, but these results do not agree quan-titatively with the single crystal work. The corresponding absorption results by these workers (shown in *Figure 15.4*) are therefore considered less reliable than those of Hall (1956). However the absorption measured by this worker seems rather small in comparison with the refractive index. From the inter-relation of the optical constants (see section 2.4), we expect the long wavelength index to be given by:

$$n_0 - 1 = \frac{2}{\pi} \int_0^\infty k \, d\omega/\omega = \frac{2}{\pi} \int_0^\infty k \, d\lambda/\lambda$$

Integrating the data of *Figure 15.4* from 0.21 μ to ∞, the expression gives only 0.33, whereas $n_0 - 1 = 1.3$. Thus the absorption shown in *Figure 15.4* is only a quarter of the total required.

The low frequency dielectric constant is $\varepsilon_0 = 11.6 \pm 1.5$ (Kroger *et al.*, 1954).

15.6.3 *Photoconductivity*

Cadmium sulphide is primarily photoconductive only in the short wavelength visible region, but specimens prepared in various ways have been shown to be sensitive to ultra-violet, infra-red and x-radiation, and to gamma rays and alpha-particles (Frerichs, 1947; Dunlap, 1957; Broser, 1957).

Even in pure single crystals the spectral distribution of photo-sensitivity can be varied markedly by changing the surface con-ditions. Most commonly the response is a sharp peak at 0.51 μ, i.e. near the absorption edge, with a steep fall at long wavelengths as the intrinsic absorption drops, and a fairly rapid fall at short wavelengths characteristic of high surface recombination (see section 4.2.3). Bube (1953) has shown that by making measure-ments in a very good vacuum the photosensitivity can be kept high at short wavelengths. Bube's results for a rough vacuum with water vapour present, and for a high vacuum, are shown in *Figure 15.7*, curves *a* and *b*. For curve *b*, $\lambda_{1/2} = 0.51$ μ.

By observing the photosensitivity to pulsed light in the presence of steady background illumination Lashkarev *et al.* (1952) have shown that the quantum efficiency is constant from 2400 to 4000 Å. Addition of oxygen gives increased sensitivity and absorption in the near infra-red region, Reynolds *et al.* (1955). Lambe (1955) has concluded that the shape of the spectral sensitivity curve can be explained .by the dependence of carrier lifetime on the wavelength, i.e. the recombination rate depends on the absorption constant or depth of penetration of the radiation.

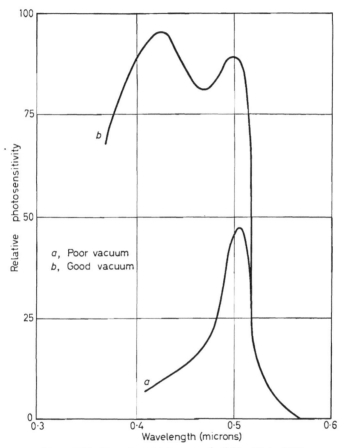

Figure 15.7. Spectral sensitivity curves for CdS. (*Bube,* 1953; courtesy *American Institute of Physics*)

Gobrecht and Bartschat (1953) have measured the photosensitivity with polarized light and obtained different spectral response curves for different crystal orientations. With the electric vector in the radiation parallel to the c axis the photocurrent peaked

219

at 5110 Å, while for radiation perpendicular to the axis the peak was at 5144 Å. This difference of 34 Å between the peaks of photoconductivity is virtually the same as the separation between the two absorption edges (section 16.6.1).

Klick (1953) has given spectral sensitivity curves for crystals at various temperatures, and finds that whereas at room temperatures the curve is sharply peaked, at 4°K the sensitivity does not fall for wavelengths below the absorption edge. Between 77°K and 300°K, the shift of the $\lambda_{1/2}$ point on Klick's curves is -8×10^{-4} eV/°C. Other values for the shift quoted by Bube (1955a) are mostly in the range -4 to -6×10^{-4} eV/°C. These values are comparable with those for the shift for the absorption edge discussed in section (16.6.1).

At 77°K Gross, Kaplyanskii and Novikov (1956) have observed a number of peaks in the photoconductivity spectra of CdS crystals. These peaks coincide with the absorption lines described earlier (section 16.6.1). The similarities and differences in the spectral dependence of absorption, photoconduction, excitation and emission have been clearly demonstrated by Klick (1953).

Sommers et al. (1956) have measured the P.E.M. short circuit current in CdS crystals. From equations (4.55) or (4.71) the square of this current is seen to be proportional to (lifetime) $\times$ (mobility)^3, for which expression these workers obtained the value 1 cm^6 V^{-3} sec^{-2}. Mobilities in CdS are not well known so that an accurate value of lifetime cannot be obtained. Roughly, however, $\mu \sim 100$ cm^2/V.sec, giving $\tau \sim 1$ μsec. The electron lifetime is always much greater than the hole lifetime (Bube, 1955a). In extreme cases the electron lifetime can reach 10^{-3} secs while the hole lifetime is 10^{-10} sec. Diemer and Hoogenstraaten (1957) and Balkanski (1958) suggest that exciton diffusion may affect P.E.M. and P.C. results.

The meaning of lifetimes, and the complex recombination and trapping phenomena which occur in CdS, are discussed in detail by Rose (1955, 1957). This author emphasizes that potential barriers are not necessary to explain the large photo-effects and the effective quantum efficiencies much greater than unity. The characteristic of CdS which is primarily responsible for the large photo-effects is the very small recombination cross section for electrons which its impurity centres can have after capturing a hole. Values as low as 10^{-24} cm^2 have been estimated from experimental data—possibly signifying capture in the face of a repulsive Coulomb field (Rose, 1957).

A more quantitative treatment of Rose's ideas has been given by

220

Bube (1957), who has been able to give a satisfactory explanation of super-linear response, temperature dependence of photocurrent and variations in response time for both CdS and CdSe photo-conductors. For CdSe this treatment gives for the important 'sensitizing' centres a capture cross section for electrons which is only 10^{-6} that for holes. For CdS these levels will be $\sim 1 \cdot 0$ eV above the valence band.

Photo-voltaic measurements made by Reynolds and Czyzak (1954) using crystals with a non-ohmic contact, demonstrate that significant electrical power outputs can be obtained. Photo-emissive measurements are reported by Shuba (1956). Orthuber and Ullery (1954) have shown that high gain image intensifiers can be made using CdS layers.

15.7 CADMIUM TELLURIDE

Cadmium telluride has a relatively low melting point (1040°C), which simplifies the problem of preparing single crystals. It shows both p-type and n-type conductivity, in contrast to CdS and CdSe. Carrier concentrations can vary from $<10^{14}$ to $>10^{19}$ cm^{-3}. According to Kroger and de Nobel (1955), who used crystals grown from the melt in a cadmium atmosphere, the mobility of electrons is 600 cm^2/V.sec, and that of holes is 50 cm^2/V.sec.

The dielectric constant is 11 at 77°K (de Nobel and Hofman, 1956). CdTe has a cubic lattice—unlike the sulphide and selenide—so birefringence will be absent.

15.7.1 Activation energy

From conductivity–temperature plots, Appel and Lautz (1954) obtained activation energies of $1 \cdot 43$ to $1 \cdot 57$ eV for the intrinsic range. Spectral sensitivity curves measured by Frericks (1947) show a sharp peak at $0 \cdot 79$ μ ($1 \cdot 57$ eV) and $\lambda_{1/2} = 0 \cdot 83$ μ ($1 \cdot 49$ eV). Miyasawa and Sugaike (1954) give $1 \cdot 47$ eV and Bube (1955b) finds $1 \cdot 41$ eV from photoconductive measurements.

Absorption edge measurements by Jenny and Bube (1954) indicate $1 \cdot 42$–$1 \cdot 45$ eV, while van Doorn and de Nobel (1956) give $1 \cdot 51$ eV for the onset of transmission through a $0 \cdot 2$ mm sample. These latter workers have measured the shift of the edge from liquid nitrogen temperature to ~ 1000°K, and also the shift of the photo-response from 300–500°K. They find that the shift varies from $-2 \cdot 3 \times 10^{-4}$ eV/°C at 77°K to $-5 \cdot 4 \times 10^{-4}$ eV/°C at 800°K. These workers also found that by biassing a p–n junction in the forward direction, emission of radiation was obtained. The radiation had a maximum intensity at $1 \cdot 40$ eV. As discussed in

221

section 6·2, this emitted radiation would be expected to have its peak at somewhat longer wavelengths than the photoconductive $\lambda_{1/2}$ point.

Pensak (1958) and Goldstein (1958) have described CdTe films which produce photo-voltages as large as 100 V. Cadmium telluride may be a useful material for solar batteries since its energy gap is near to the optimum calculated in section 4.6.

Transmission measurements on mixtures of CdTe and HgTe have been reported by Lawson *et al.* (1958). The edge moves progressively to longer wavelengths with increasing HgTe content, reaching 15 μ for 90 per cent HgTe. Photoconductivity in some mixtures extends to 13 μ.

CHAPTER 16

INDIUM ANTIMONIDE AND OTHER
III–V COMPOUNDS

16.1 Introduction

THE so-called 'intermetallic' compounds of elements of groups
III and V of the periodic table form a particularly interesting
series of semi-conductors, and since the first publications concerning
them by Welker at the end of 1952, they have been studied
intensively. Already the literature on these materials is extensive,
and they have been the subject of three recent review articles,
namely (1) Pincherle and Radcliffe (1956), (2) Welker and Weiss
(1956), and (3) Cunnell and Saker (1957). There is also a collection
of papers in Vol. 2, No. 2 of the Journal of Electronics, 1955.

As these reviews are up to date and generally available, the
present chapter will give only a brief survey of the properties of
the whole group, and then a detailed account of the most interesting
one of these compounds—and perhaps the most interesting of all
semi-conductors—namely indium antimonide.

16.2 General survey

The compounds of interest are the nine combinations of Al, Ga
and In with P, As and Sb. All crystallize in the zinc blende lattice,
with interatomic spacings as shown in *Table 16.1*.

Table 16.1. *Interatomic Distances for* III-V *Compounds* (Å)

	P	As	Sb
Al	2·36	2·44	2·63
Ga	2·36	2·44	2·63
In	2·54	2·62	2·80

It will be seen that the lattice spacings of the group IV elements,
namely:

Si 2·35 Å,   Ge 2·44 Å,   Gray Sn 2·80 Å,

are very close to those for the compounds of elements of the same
row of the periodic table. The coefficient of linear thermal expansion is near $5·5 \times 10^{-6}$ per °C for all the compounds (Welker
and Weiss, 1956).

223

The results of transmission measurements by Welker and Weiss (1956) and co-workers are given in *Figure 16.1*. For the seven compounds shown, the absorption edges cover the range 0·55 μ to 7·5 μ. All the absorption edges move to longer wavelengths on

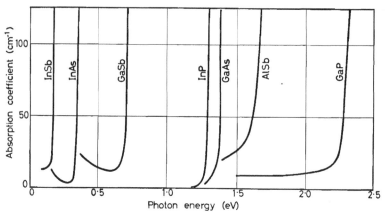

Figure 16.1. Absorption spectra of III–V compounds. (Welker and Weiss, 1956; courtesy Academic Press, New York)

heating. The values of the energy gaps determined by both optical and electrical measurements, and the temperature dependence of the gaps, are summarized in *Table 16.2*.

Table 16.2. Activation Energies of Intermetallic Compounds

Compound	E(Opt.) 300°K eV	E(Elect.) 0°K eV	dE/dT eV/°C	n
InSb	0·18	0·26	$-2\cdot9 \times 10^{-4}$	3·96
InAs	0·33	0·45	$-3\cdot5 \times 10^{-4}$	3·5
InP	1·26	1·34	$-4\cdot6 \times 10^{-4}$	3·1
GaSb	0·70	0·82	$-4\cdot1 \times 10^{-4}$	3·8
GaAs	1·35	1·5	$-4\cdot9 \times 10^{-4}$	3·4
GaP	2·24		$-5\cdot4 \times 10^{-4}$	2·9
AlSb	1·6	1·65	$-3\cdot5 \times 10^{-4}$	3·2
AlAs	2·16			
AlP	3·1			3·4

It will be observed from the table that for compounds of any given element (either group III or group V) the energy gap decreases with increasing atomic weight of the other element of the combination. It also appears that the spectral shift increases progressively through the series: antimonide, arsenide, phosphide. By the

224

use of mixed crystals of InAs/InP and GaAs/GaP, a wide variety of activation energies can be achieved (Welker and Weiss, 1956).

Also shown in *Table 16.2* are the values of the refractive index. As expected from the polarizability of the ions, (Moss, 1950) the index increases through the series: phosphide, arsenide, antimonide.

It will be seen that, as expected from equation (3.25), $n^4 E$ is fairly constant except for InAs and InSb, with values close to those for the group IV elements.

Photo-effects have been observed in GaAs, AlSb, GaSb, InAs and InP. The first two compounds are of interest for use in solar batteries, and a promising unit of GaAs has already been made by Jenny *et al.* (1956). InAs has been shown by Talley and Enright (1954), Hilsum (1957) and Dixon (1957) to be a promising detector for the near infra-red region—using either photoconductive, photo-electro-magnetic, or p–n junction voltaic effects. Lukes (1956) has found a quadratic temperature dependence for the photoconductive limit of GaSb. Photo-voltaic effects in InP have been studied by Reynolds *et al.* (1958).

Braunstein (1955) has observed recombination radiation from GaSb and GaAs. The peak emissions at 300°K are at 0·625 eV and 1·1 eV respectively, i.e. at somewhat lower photon energies than those shown for the energy gap in *Table 16.2*. This behaviour is as expected from the theoretical treatment of section 6.2.

Long wavelength peaks in reflectivity have been observed for five of the compounds by Picus *et al.* (1957). The wavelengths of these peaks are shown in *Table 16.3*. The value for InSb by Yoshinaga (1955) is included.

Table 16.3. Restrahlen Wavelengths (microns)

InP	InAs	GaAs	GaSb	AlSb	InSb
32	41	36	40·5	31	54·6

It will be seen that for compounds involving similar atomic weights, which consequently have the same reduced mass, (e.g. InAs and SbGa) the wavelengths are approximately equal. For compounds with different reduced masses however, the expected dependence of Restrahlen wavelength on atomic weights (section 3.10) is not obtained.

The electrical and optical properties of alloys of III–V compounds with group IV elements have been reported by Kolm *et al.* (1957).

225

With the exception of Ge in GaAs, addition of the group IV element always increases the activation energy.

16.3   GENERAL PROPERTIES OF InSb

Since the first measurements were made on this material in 1952 it has been the subject of widespread and intensive research work. Many properties of this compound have extreme values in comparison with other known semi-conductors, and already it has been found to show several effects which are not detectable in other materials. It is a promising material for use in a variety of technological applications (Moss, 1958b).

Many of the interesting properties of InSb stem from the following parameters: (1) small intrinsic energy gap, 0·18 eV at room temperature, and (2) very small effective mass of electrons, only 1·5 per cent of the mass of a free electron. As a result of (1) InSb has useful properties for infra-red detectors and filters*. Also, in this part of the spectrum high resolution in photon energy can readily be achieved. For example, a change in the energy gap of only 10^{-3} eV gives the relatively large change in wavelength of 0·05 μ. The small mass is mainly responsible for the very high mobility of InSb—70,000 cm²/V.sec—at room temperature. This makes the material twenty times more sensitive to first order magnetic field effects than Ge for example, and four hundred times more sensitive to second order effects.

The melting point of the compound is low, so that zone melting and crystal growing are relatively easy. It is now normal laboratory routine to produce material which has a nett extrinsic carrier concentration only 10 per cent of the intrinsic room temperature carrier concentration, and specimens with ten times this purity are not too uncommon.

In such pure material lattice scattering dominates the mobilities at room temperature and somewhat below. Hrostowski et al. (1955) conclude that the lattice scattering mobilities vary as

$$\mu_e \propto T^{-1·68}, \qquad \mu_h \propto T^{-2·1}$$

At 78°K, an electron mobility of 500,000 cm²/V.sec was observed by Harman et al. (1955). The mobility of holes is not well known. The best values are 750 cm² by Hilsum and Barrie (1958) and that determined by Goodwin (1956) from photo-effects, namely 800 cm²/V.sec, at room temperature.

The intrinsic carrier concentration at 295°K is

$$n_i = 1·6 \times 10^{16} \text{ cm}^{-3}$$

* Smith and Moss (1958) describe bloomed InSb filters.

226

Due to the large mobility ratio, maximum resistivity is not obtained with intrinsic material, but with

$$n \doteqdot 2 \times 10^{15} \, \text{cm}^{-3}, \qquad p \doteqdot 1 \cdot 3 \times 10^{17} \, \text{cm}^{-3}$$

Recent publications on the electrical properties of p-type InSb are those by Fukuroi and Yamanouchi (1957), Frederikse and Hosler (1957) and Hilsum and Barrie (1958).

16.4 BAND STRUCTURE OF InSb

Early speculations by Herman (1955b) were that both conduction band minimum and valence band maximum were spherical energy surfaces at $\tilde{k}000$. The isotropic behaviour of magnetoresistance (Pearson and Tanenbaum, 1953) and cyclotron resonance (Dresselhaus et al., 1955b) support this view. There is also some evidence from cyclotron resonance for two hole masses, corresponding to two valence bands. From the helium temperature radio frequency cyclotron resonance work, the electron mass is $m_e = 0 \cdot 013m$, and one of the hole masses is $m_h = 0 \cdot 18m$. It does not seem to be established at present whether the mass of the other hole is greater or less than the above. As in Ge, there will be a third valence band separated from the above two by spin orbit splitting. Since the splitting is estimated to be $\sim 1 \cdot 0$ eV (Kane, 1957), this band will have a negligible effect on the properties.

After some uncertainties regarding the position and number of valence band maxima, it seems now to be established from the interpretation of mechanical, optical and electrical data (see Potter, 1957a, or Tuzzolino, 1957,8; Dumke, 1957b; Frederikse and Hosler, 1957, respectively) that both extrema are at, or extremely close to, $\tilde{k}000$.

The band structure has been calculated by Kane (1957). This work was not a complete calculation from the basic atomic wave functions, but assumed at the outset that both band extrema were at $\tilde{k}000$. Also the measured values for the energy gap and effective masses were taken as known data, and then the detailed shape of the energy surfaces computed. The most noteworthy feature is that the conduction band is highly non-parabolic, so that the effective mass increases with energy.

The validity of Kane's treatment is confirmed by the good agreement between his calculated values of absorption coefficient and the measured values described in the next section. He assumes that the cyclotron resonance hole mass of $0 \cdot 18m$ corresponds to the 'heavy' holes of the uppermost valence band.

Transitions between the valence bands have not so far been

227

reported, although such absorption has been observed in InAs (Stern and Talley, 1957).

16.5 SHORT WAVELENGTH ABSORPTION IN InSb AND ACTIVATION ENERGY

Measurements at room temperature of the absorption coefficient by direct transmission through very thin single crystal specimens of intrinsic material by Moss, Smith and Hawkins (1957) are shown in *Figure 3.3*.

The main absorption edge is seen to be near 7 μ. The wavelength of maximum rate of change of K is 6·94 μ, or 0·179 eV. Also shown in *Figure 3.3* are the values of K calculated by Kane (1957) from his band structure theory, on the basis of direct transitions. Since in this calculation the only adjustable parameter was the position of the edge, the agreement in absolute magnitude at short wavelengths is good.

At wavelengths below 4 μ it is found empirically that K increases inversely as λ, reaching $1·6 \times 10^4$ cm^{-1} at 1·5 μ. This is a much higher absorption level than that at which the onset of vertical transitions in Ge and Si shows a secondary absorption edge. This fact indicates that there is only the one absorption edge, and that the conduction band minimum is therefore vertically above the valence band maximum.

Interpretations to the contrary—in terms of non-vertical transitions involving one, two or four acoustical mode phonons and valence band maxima away from $\bar{k}000$—have been made by Roberts and Quarrington (1955), Blount *et al.* (1956) and Potter (1956*a*).

However, the most recent work on this subject by Dumke (1957*b*) shows that the absorption edge can best be explained by a model with both extrema at $\bar{k}000$, where, at energies slightly below those of direct transitions, there is absorption due to virtual transitions to intermediate states in the conduction band, followed by scattering by optical mode phonons of energy appropriate to the Restrahlen wavelength. (See section 16.7.) Dumke's results are compared with the experimental results of Roberts and Quarrington (1955) in *Figure 16.2*. For temperatures of 77°K and below the *shape* of the absorption curve is independent of temperature, and further cooling merely shifts it to shorter wavelengths.

A particularly precise value for the energy gap has been obtained from a study of the oscillatory magneto-absorption effect by Zwerdling, Lax and Roth (1957) who have extended the original work of Burstein and Picus (1957). Using fields of 20,000–37,000 G and specimens of about 7 microns thick these workers were able to locate four absorption minima in the wavelength range 5–7 μ. On plotting the photon energies of these minima against magnetic field it is found that on extrapolation to zero field they all yield the same photon energy. This is the energy gap, and its value is $0{\cdot}180 \pm 0{\cdot}002$ eV at 298°K, confirming the value obtained from

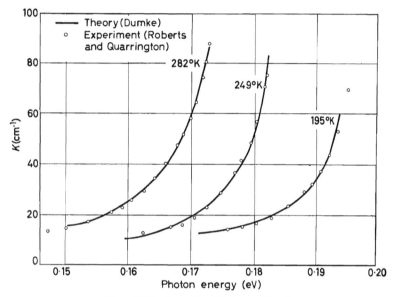

Figure 16.2. Absorption of indium antimonide

the maximum slope of the absorption curve. From the magneto-absorption work the effective mass was found to be $m_e = 0{\cdot}014m$, with negligible anisotropy. Boyle and Brailsford (1957) obtained $m_e = 0{\cdot}0146m$ from magneto-absorption of bound carriers.

A shift of the absorption edge in high magnetic fields has been detected by Burstein, Picus and Gebbie (1956a). Using fields up to 58,000 G they found that for a constant level of transmission there was a change of $2{\cdot}3 \times 10^{-7}$ eV/G in the photon energy. This is only ∼60 per cent of the expected shift given by

$$dE/dB = h/4\pi m^* \text{ eV per weber m}^{-2}$$

possibly because of a variation in absorption magnitude due to change of electron populations caused by the field (Landsberg, 1958).

16.5.1   Pressure and temperature dependence of activation energy of Insb

Because of the technical difficulties of providing a window material transparent at ~8 μ which will stand high pressures, no measurements of the pressure dependence of the absorption edge have been reported. However work has been carried out on the resistivity and Hall constant by Long (1955), Keyes (1955) and Long and Miller (1955). The average increase of energy gap with pressure is:

$$dE/dp = 15 \times 10^{-6} \text{ eV/kg cm}^{-2}$$

The elastic constants of InSb measured by Potter (1956*b*) and de Vaux and Pizzarello (1956) give values of compressibility of

$$\chi = 2 \cdot 3 \times 10^{-12} \text{ and } 2 \cdot 1 \times 10^{-12} \text{ per dyne cm}^{-2} \text{ respectively}$$

or

$$\chi = 2 \cdot 2 \times 10^{-6} \text{ per kg cm}^{-2}$$

Thus the equivalent shift per unit strain is

$$dE/(dV/V) = -6 \cdot 8 \text{ eV}$$

Hence from equation (3.18),

$$\pm C_e \pm C_h = -10 \text{ eV}$$

and the equivalent temperature dependence due to dilatation (equation 3.20) is

$$(dE/dT)_d = -1 \cdot 1 \times 10^{-4} \text{ eV/}°\text{C}$$

The temperature dependence of the energy gap has been measured in a variety of ways. From a study of the absorption edge, Roberts and Quarrington (1955) found that the shift was linear with temperature above 100°K. Results for temperatures greater than this are summarized in *Table 16.4*.

Table 16.4. Temperature Dependence of Energy Gap of InSb

Method	Temperature °K	Shift eV/°C	Observer
Absorption	100–290	$-2 \cdot 9 \times 10^{-4}$	Roberts–Quarrington
Absorption	77–300	$-4 \cdot 0 \times 10^{-4}$	Tanenbaum–Briggs
Absorption	100–500	$-2 \cdot 8 \times 10^{-4}$	Fan
Absorption	100–500	$-2 \cdot 6 \times 10^{-4}$	Austin–McClymont
Absorption	100–500	$-2 \cdot 6 \times 10^{-4}$	Oswald
Absorption	300–455	$-3 \cdot 9 \times 10^{-4}$	Moss
Conductivity	200–500	$-3 \cdot 9 \times 10^{-4}$	Hrostowski, *et al.*
Photo-effects	175–231	$-3 \cdot 5 \times 10^{-4}$	Tauc–Abraham
P.E.M.	100–200	$-3 \cdot 6 \times 10^{-4}$	Kurnick–Zitter
Emissivity	310–400	$-2 \cdot 9 \times 10^{-4}$	Moss
		Average	
		$-3 \cdot 3 \times 10^{-4}$	

In conjunction with the dilatation contribution of $-1 \cdot 1 \times 10^{-4}$ eV/°C, this average value gives a lattice broadening contribution

$$(\mathrm{d}E/\mathrm{d}T)_b = -2 \cdot 2 \times 10^{-4} \text{ eV/°C}$$

If the deformation potential theory of mobility, (equation 3.19) is assumed to be applicable, values of C_e and C_h may be obtained. From the elastic constants (Potter, 1956b)

$$\rho u^2 = 0 \cdot 79 \times 10^{12} \text{ cgs units}$$

Using

$$\mu_e = 70,000, \qquad \mu_h = 800, \qquad m_e = 0 \cdot 015 m, \; m_h = 0 \cdot 18 m$$

we obtain from equation (3.19)

$$C_h = 32 \text{ eV}, \qquad C_e = 76 \text{ eV}$$

Equations (3.20) and (3.21) for the shift contributions can be satisfied by putting

$$C_h = 32 \text{ eV}, \qquad C_e = 42 \text{ eV}$$

These values show that the two extrema move in the same direction on compression, so that the pressure effect is given by the difference of C_e and C_h. The fact that the value of C_e obtained from the shift is less than that from the mobility is perhaps explained by the statement of Ehrenreich (1957) that although the electron mobility is very high it is difficult to see theoretically why it is not much higher. If only about one-third of the scattering was attributable to deformation potential effects, so that a mobility $\sim$200,000 cm²/V.sec could be used in equation (3.19), then a value of $C_e = 42$ eV would be obtained.

16.6 CONCENTRATION DEPENDENCE OF ACTIVATION ENERGY

It was first noticed by Tanenbaum and Briggs (1953) that the position of the absorption edge in n-type InSb was strongly dependent on the impurity concentration. The transmission of specimens doped with various amounts of tellurium (Hrostowski et al., 1954) are shown in *Figure 16.3*. With extrinsic carrier concentrations varying from 5×10^{15} cm⁻³ to $\sim 10^{19}$ cm⁻³ the absorption edge moves from 0·18 eV to $\sim$0·6 eV.

The effect has been explained by Burstein (1954) and Moss (1954a) by postulating that the lower states of the conduction band are filled progressively by the extrinsic electrons, so that absorption transitions can only take place to the higher, empty, conduction band states.

The theory for a simple parabolic conduction band (i.e. constant effective mass) is given in section 3.4, where it is shown that the shift of the edge is given by

$$\Delta E = 4 \times 10^{-15} \, (m/m_e) N^{2/3} \tag{16.1}$$

However if the band is non-parabolic with a linear dependence of E on $\tilde{k}$ for example, then we obtain:

$$\Delta E \propto N^{1/3} \tag{16.2}$$

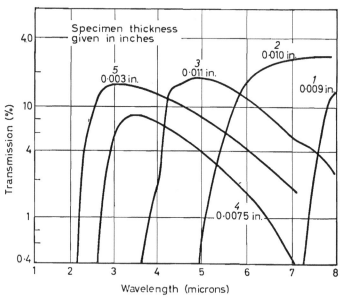

Figure 16.3. Transmission of n-type InSb. (Hrostowski et al., 1954; courtesy American Physical Society)

In order to illustrate the difference between equations (16.1) and (16.2) the available experimental data (Kaiser and Fan, 1955; Hrostowski et al., 1954; Breckenridge et al., 1954) have been plotted against both $N^{1/3}$ and $N^{2/3}$ in Figure 16.4. There is considerable scatter of the points, but it may be seen that the trend of one plot is generally convex and the other concave to the axis, so that the overall variation may be taken as roughly $\Delta E \propto N^{1/2}$. Alternatively the slope of the plot against $N^{2/3}$ can be taken as giving the reciprocal of the effective mass, so that this mass increases with increasing N.

This variation of effective electron mass with energy has been confirmed directly by cyclotron resonance measurements at infrared wavelengths. Such measurements were first made by Burstein, Picus and Gebbie (1956a) and were then extended by Keyes *et al.* (1956). By use of pulsed magnetic fields up to 300,000 G these latter workers were able to make measurements over a significant range of energies. Wallis (1958) has discussed the theory of cyclotron

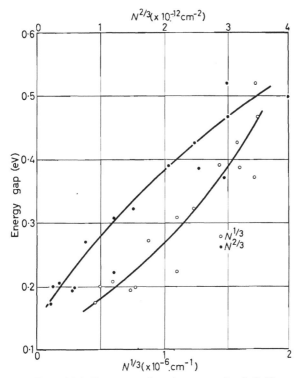

Figure 16.4. Energy gap—impurity concentration for InSb

resonance in non-parabolic energy bands, and has pointed out that the above results may be somewhat in error as the resonance peaks were determined by varying magnetic field instead of frequency, so that changes in carrier concentration with field would affect the response curves. However the effect should not be very serious.

The results obtained are plotted against the energy corresponding to the magnetic field (i.e. $heB/2\pi m^*$) in *Figure 16.5*. They indicate a greater slope than the theoretical line for m^* derived from the

233

band theory of Kane (1957). Agreement would be better if the mass were plotted against $1\cdot5\hbar\omega_c$—i.e. the upper instead of mean level of the most probable transition.

Also included in *Figure 16.5* are the oscillatory magneto-optical values of effective mass, and the radio frequency cyclotron resonance mass, and results discussed in sections 16.8 and 16.10.

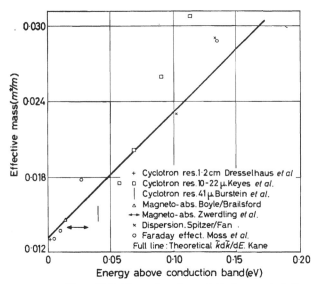

Figure 16.5. Energy dependence of effective mass in InSb

16.7 LATTICE ABSORPTION (RESTRAHLEN BAND)

The reflectivity of pure InSb crystals in the region of the Restrahlen band has been measured by Yoshinaga and Oetjen (1956). The reflectivity falls steadily as the wavelength is increased beyond 15 μ to reach a minimum at 49·8 μ (at 25°C). There is then a rapid rise to a maximum reflectivity of 83 per cent at 54·6 μ. This reflectivity figure is rather higher, and presumably more accurate than the 78 per cent quoted earlier by Yoshinaga (1955).

These measured values can be used to derive the parameters of the equivalent classical oscillator, using the theory of section 2.3. From equations (2.23*b*), (2.21) and (2.25*b*) we have:

$$(1 + R)/(1 - R)_{\text{max}} = 1\cdot3\rho, \qquad \text{or } \rho = 8$$

$$\omega_{\text{max}} - \omega_0 = 1\cdot12g, \qquad \omega_{\text{min}} - \omega_0 = 4\cdot46g$$

234

Hence
$$g \equiv 5 \cdot 4 \text{ cm}^{-1}, \qquad \omega_0 \equiv 177 \cdot 6 \text{ cm}^{-1}$$
and
$$Ne^2/m\varepsilon_0 \equiv 12 \times 10^4 \text{ cm}^{-2}$$

From these values, the magnitude of the reflection minimum and the calculated reflectivity at 40 μ and 60 μ are all in good agreement with the experimental values. Hence it is concluded that the assumption of a single classical oscillator to represent the Restrahlen band is a good approximation.

The parameters calculated above represent a very narrow and intense absorption band, with a maximum absorption at 56·4 μ given by $2nk = 128$, or $K = 1 \cdot 7 \times 10^4 \text{ cm}^{-1}$. The nett contribution to the low frequency dielectric constant by the band is:

$$Ne^2/\varepsilon_0 m\omega_0{}^2 = 3 \cdot 7$$

which indicates a considerable degree of ionicity. The effective value of ionic charge may be obtained from this value. According to Burstein and Egli (1955) the required relation is,

$$\Delta\varepsilon = (n_1{}^2 + 2)^2 Ne^2/9\mu\varepsilon_0\omega_0{}^2 \qquad (16.3)$$

where the reduced mass $\mu = 10^{-25}$ kg, the density of 'molecules' $N = 1 \cdot 5 \times 10^{28} \text{m}^{-3}$, and the optical dielectric constant $n_1{}^2 = 15 \cdot 9$. Thus $e = 0 \cdot 5$ free electron charges. This value agrees with the charge deduced by Ehrenreich (1957) from the mobility data. From the elastic moduli, Potter (1957*b*) obtains $e = 0 \cdot 7$ free charges.

Weak lattice absorption bands at 28·2 μ (i.e. the Restrahlen overtone) and at 30 μ have been reported by Spitzer and Fan (1955).

16.8 REFRACTIVE INDEX AND DISPERSION OF InSb

The refractive index has been measured by an interference method over the wavelength range 7–20 μ by Moss, Smith and Hawkins (1957). As pointed out in section 1.5, it is not possible to determine the true value of refractive index from the *separation* of fringes if there is a linear term in the dispersion. In order to avoid such an error in this work some of the specimens used were made so thin that the actual fringe number could be determined unambiguously.

The results obtained for pure single crystal material are shown plotted against wavelength (on linear scales) in *Figure 16.6*. Over the wavelength range of 9–20 μ the spacing of the fringes was virtually constant, but this does not mean that the index is constant over this range, for as the figure shows there is strong dispersion with almost linear variation of n with λ.

235

The computed refractive index curve in *Figure 16.6* is synthesized from four separate contributions. These are:

(*1*) *Tail band absorption*—The measured absorption in the long wavelength tail of the main absorption band has been plotted in *Figure 3.3*. The dispersion due to this band is obtained from the theory of the inter-relation of the optical constants in section 2.4. Graphical integrations of suitable functions of the absorption constant were carried out for each desired wavelength. The results showed that this was the most important contributor to the dispersion in the 5–10 μ region, and that it was responsible for the peak shown in *Figure 16.6*.

(*2*) *Short wave band absorption*—As no data were available on the absorption below 1·5 μ, it was assumed to be that of a simple classical oscillator. The incremental contribution to the dielectric constant of this band was found to be

$$\Delta\varepsilon = 3\cdot5/\lambda^2$$

This is a very small contribution in the wavelength range required, so that its accurate assessment is not important.

(*3*) *Free carrier absorption*—The free carrier dispersion is given by equation (2.30a). As $\omega \gg g$ and $n^2 \gg k^2$ in this wavelength region, we have

$$n^2 - \varepsilon = -Ne^2/m^*\varepsilon_0\omega^2$$

With $N = 1\cdot6 \times 10^{22}\text{m}^{-3}$ and $m^*/m = 0\cdot015$ we obtain:

$$\Delta\varepsilon = 9\cdot2 \times 10^{-4}\lambda^2$$

(*4*) *Restrahlen band*—The parameters of the Restrahlen absorption band were calculated in section 16.7. In the required wavelength range, $\omega \gg g$, so that the dispersion is given by

$$\Delta\varepsilon = Ne^2/m\varepsilon_0(\omega_0{}^2 - \omega^2) = -4\lambda^2/(\lambda_0{}^2 - \lambda^2)$$

where $\lambda_0 = 56\cdot4$ μ. This term is somewhat larger than the free carrier term.

As shown by the dashed curve of *Figure 16.6* the sum of these four contributions agrees well with the measured values over the whole range. The treatment shows clearly how a substantially linear dependence of n of λ is built up from a combination of conventional dispersion mechanisms. InSb is perhaps unique in having these four dispersion terms overlapping to such a marked degree, but it is possible that other materials could show the effects to lesser extent. It has been pointed out in section 10.4 that there may be a small linear term in the dispersion of Ge.

As there is no wavelength region where the refractive index

236

becomes constant it is not possible to determine a 'long wavelength refractive index' from the experimental results alone. Study of the individual contributions, however, shows that the value which the index would have at long wavelengths if there were no free carrier or Restrahlen dispersion occurs at 9·8 μ, and its value is $n_{9·8} = 3·96$. Yoshinaga and Oetjen (1956) conclude from their

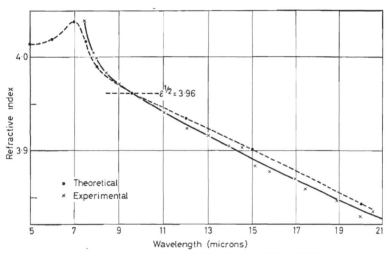

Figure 16.6. Refractive index of InSb. *(Moss, 1957b; courtesy Physical Society, London)*

long wavelength measurements at different temperatures that the undispersed value of the optical dielectric constant increases with temperature.

The dispersion due to free carriers in impure material has been studied by Spitzer and Fan (1957b). These workers point out that the theoretical expression for the frequency dependence of the dielectric constant (equation 2·30a) does not involve the collision frequency (g) if $\omega^2 \gg g^2$, unlike equation (2.30b) for the absorption term. Thus they expect the real part to vary accurately with (wavelength)^2, even if g is energy dependent.

The above behaviour was verified experimentally for both p and n type material, although the absorption coefficient did not in general vary as λ^2. In particular, for p type material K was virtually constant over the range of measurements (i.e. 6 μ to 12 μ). From the λ^2 dependence of the dielectric constant an effective mass $m_h = 0·2m$ was deduced. For n-type material, K was approximately

237

proportional to λ^2 at low levels, but increased more rapidly with λ at high levels. Most of this discrepancy however is removed if account is taken of the change in refractive index, for as shown by equation (2.32) it is nK which should vary as λ^2. Replotting the data of Spitzer and Fan (1957b) in this manner (*Figure 2.3b*) shows that the dependence is approximately as λ^2, so that the collision time $(1/g)$ must be almost independent of energy in the range studied.

It is pointed out that the effective mass for the non-parabolic band which determines the optical properties is not given by: $\hbar^2 m^* = d^2 E/d\bar{k}^2$ but is defined by equation (5.17). These two expressions are of course the same for a simple parabolic band, but not otherwise. These effective masses were determined for n-type samples of different purity. From the known carrier concentrations the height to which the conduction band would be filled has been calculated, and the effective mass is plotted against this energy in *Figure 16.5*.

16.9 RECOMBINATION RADIATION AND EMISSIVITY

The measured spectral distribution of recombination radiation from InSb is shown in *Figure 6.2*. The results are seen to agree reasonably well with theoretical expectations, as discussed in section 6.2.

The radiation lifetime has been calculated from equations (6.3) and (6.7) using the data of *Figure 3.3*. The value found is $\tau_R = 0.79$ μsec. Typical measured values of τ in good quality crystals are $\sim 4 \times 10^{-8}$ sec, so that it appears that only ~ 5 per cent of the recombinations are radiative. From the absolute intensity of the recombination radiation it was estimated in section 6.2 that this figure was ~ 20 per cent. Landsberg (1957) has suggested that this discrepancy could be explained if the short wavelength light used to generate the excess carriers produced more than one electron-hole pair per photon. As discussed in Chapter 10, such an enhanced quantum efficiency was first found in germanium by Koc (1957) and later in InSb by Tauc (1958).

The expected dependence of τ_R on impurity concentration has been calculated by Landsberg and Moss (1956), and is shown in by Tauc (1958) *Figure 6.1*.

Measurements of the emissivity of InSb single crystals by Moss and Hawkins (1958a) are shown in *Figure 16.7*. It will be seen that measurements can be made at little above room temperature. A plot of photon energy for half maximum emissivity against temperature shows that the shift of the edge is -2.9×10^{-4} eV/°C. The absolute value of the emissivity at short wavelengths was measured using an InSb detector. This detector automatically responds to almost all the wavelength range of high emissivity, but is insensitive to wavelengths beyond the emission edge. The values obtained lay in the range 63–72 per cent, are in agreement with the value calculated from equation (6.26) using the available data on reflectivity.

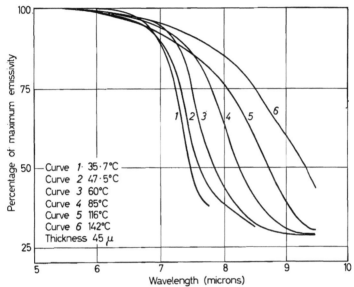

Figure 16.7. Emissivity of InSb. (*Moss and Hawkins, 1958a; courtesy Physical Society, London*)

16.10 INFRA-RED FARADAY EFFECT IN InSb

From the theory of Chapter 5, the Faraday rotation due to free carriers is related to the dispersion by equation (5.16), which may be rewritten:

$$Be/m^* = 2c\lambda\theta/(dn/d\lambda)$$

This equation shows that one can in effect measure the cyclotron resonance frequency (Be/m^*) by determining the Faraday rotation and dispersion at a given wavelength. The great attraction of this method compared with a direct determination of cyclotron resonance is that the huge magnetic fields required for infra-red measurements, or very low temperatures needed for radio measurements, are avoided. The low value of m^* for electrons in InSb makes the measurement particularly easy, but as θ can, in general, be measured quite accurately, much larger values of effective mass could also be determined in this way.

As pointed out in section 16.8, the dispersion is independent of g, and thus of the detailed scattering mechanisms at infra-red wavelengths where $\omega^2 \gg g^2$. The same condition applies to the Faraday effect, as shown by equation (5.13a).

The approximation $(Be/m)^2 \ll \omega^2$ is fulfilled for InSb in the wavelength range of 10–20 μ for all convenient fields (e.g. for 15 μ and 20,000 G, $(Be/m)^2 = \omega^2/35$). For a material with $m^* \sim m$, the approximation would hold up to 10^6 G.

Measurements of the Faraday rotation have been made by Moss and Smith (1958) on both pure and n-type InSb in the wavelength range 11–22 μ. For the n-type material the dispersion was also measured. For this material nearly all the dispersion at these wavelengths is due to free carriers, so that equation (5.16) can be applied directly to find the effective mass. The results at various wavelengths and magnetic fields give for material with 6.4×10^{17} cm^{-3} free electrons the average value $m_e = 0.029m$.

For purer material, the dispersion due to the Restrahlen band is of the same order as the free carrier dispersion, so that it is preferable to use equation (5.15). Also to ensure degeneracy it was necessary to cool to 77°K. Results for the effective mass (defined as in equation 5.17) are plotted against the position of the Fermi level in *Figure 16.5*. They are seen to be in good agreement with the theoretical value of m^* calculated from the E–$\tilde{k}$ curve given by Kane (1957).

From equations (5.15) and (2.32) it will be seen that the ratio of Faraday rotation to absorption coefficient is given simply by $\theta/K = \frac{1}{2}\mu B$. Thus the maximum rotation for a given free carrier absorption is independent of wavelength and is proportional to mobility, explaining why the effects are so large in InSb. It may be shown that the maximum product of rotation and transmission occurs for a specimen thickness $1/K$, when the loss is $e:1$ and the rotation is $\frac{1}{2}\mu B$ radians. For 10^4 G this rotation reaches 200° for intrinsic InSb.

16.11 PHOTO-EFFECTS IN InSb

Photoconductive, P.E.M. and photo-voltaic effects at both point contacts and p–n junctions have all been observed in InSb.

The spectral response of photoconductivity is shown in *Figure 4·1*. It will be seen at room temperature the sensitivity extends to $\lambda_{1/2} = 7.7$ μ. The edge moves to much shorter wavelengths at low temperatures, so that the high sensitivities which have been obtained for liquid air cooled p–n junctions by Mitchell et al. (1955), Avery et al. (1957) and Lasser et al. (1958) are offset by the relatively poor spectral response under these conditions. The main interest therefore lies in the performance without cooling.

The lifetime of the photocarriers is very short. Measurements of the magnitude of photoconductivity, P.E.M., or preferably the ratio of the two, give lifetimes up to $\sim 4 \times 10^{-8}$ sec for good quality single crystals at room temperature. The most thorough study of lifetime in InSb has been made by Wertheim (1956) who made direct measurements of the rise time of conductivity induced by bombardment with very short pulses of 700 kV electrons. The maximum lifetime occurred near 200°K and was $\sim 0\cdot5$ μsec, falling to $0\cdot15$ μsec at 250°K. After subtracting Shockley–Read type recombination from the measured results, the residual recombination was the same for all samples, and had the temperature dependence expected for radiative recombination, but the magnitude of this experimental radiative lifetime was only one-third of the theoretical τ_R.

These short lifetimes favour the P.E.M. effect (which falls off only as $\tau^{1/2}$) and it is a P.E.M. mode detector which is the most sensitive uncooled detector yet described (Hilsum and Ross, 1957).

Photoconductive and P.E.M. effects have been studied by Hilsum *et al.* (1955) and Kurnick and Zitter (1956*b*). Results by the latter workers (*Figure 4.5*) show that the magnetic field dependence of P.E.M. short circuit current given by either equation (4.84) or (4.85) can be obtained by electrolytic or mechanical polishing of the specimen surfaces respectively. Surface recombination velocities as high as 10^6 cm/sec were observed in this work.

It has been shown by the author (1954*b*, 1958*a*) that the very low effective mass of the carriers in InSb is important in giving a relatively low intrinsic carrier concentration, and hence high photosensitivity, for such a small energy gap. The high mobility ratio is also advantageous, since by doping the material to be somewhat *p*-type, the dark conductivity can be reduced considerably below that of the intrinsic material. As shown by Wertheim (1956) the carrier lifetime increases on cooling to ~ 200°K, and this results in increased photosensitivity (Goodwin, 1957). The optimum amount of cooling to be used is a compromise between gain of sensitivity and loss of spectral response, but cooling by ~ 50°C gives a considerable improvement in sensitivity for little loss in wavelength range.

Both Suits *et al.* (1956) and Oliver (1957) have found that InSb photo-elements do not show any significant amount of current noise.

THEORY OF TRANSITION LAYER AT INTERFACE OF TWO DIELECTRICS

THE experimental result that when plane parallel light is reflected from a dielectric at the Brewster angle it is elliptically polarized to a slight degree, can be explained by the presence on the surface of the dielectric of a thin transition layer, within which the dielectric constant changes continuously from its value in the first medium, ε_1, to that in the second medium, ε_2. It has been shown by Drude (1902) that the ellipticity at the Brewster angle (i.e. ratio of wave *amplitudes*) is given by

$$\bar{\rho} = \frac{\pi(\varepsilon_1 + \varepsilon_2)^{1/2}}{\lambda(\varepsilon_1 - \varepsilon_2)} \int_{x=0}^{x=l} \frac{(\varepsilon - \varepsilon_1)(\varepsilon - \varepsilon_2)}{\varepsilon} \, dx$$

where l is the layer thickness.

Consider the case for heavy flint glass (for which $\varepsilon_2 = 3$), with $\varepsilon_1 = 1$ for air as the first medium, where it is found experimentally that $\bar{\rho} = 0.03$ in the visible region. It is reasonable to assume that the dielectric constant will vary monotonically in some sigmoid fashion, and we will assume for convenience that $\varepsilon = 2 - \cos \pi x/l$. Evaluation of the integral gives $\bar{\rho} = 0.85 l/\lambda$ so that the observed value $\bar{\rho} = 0.03$ corresponds to $l/\lambda = 0.035$ or a transition layer $\sim \lambda/30$ thick.

An indication of the effect of assuming different laws for the variation of ε with x is given by taking the maximum possible value of the integral for $\bar{\rho}$, which would occur if throughout the whole of the transition layer $\varepsilon = (\varepsilon_1 \varepsilon_2)^{1/2}$. For the parameters used above this extreme case gives $l/\lambda = 0.018$ or $l = \lambda/55$. As another example, a linear variation throughout the transition layer gives $\bar{\rho} = 1.1 l/\lambda$ or $l/\lambda = 0.027$.

APPENDIX B

INTER-RELATION OF OPTICAL CONSTANTS

B.1 Derivation of the Relation Between the Real and Imaginary Parts of the Dielectric Constant

For a single classical oscillator of resonant frequency ν_0 and strength A_0 we have that at some specific frequency ν_c the imaginary part of the dielectric constant is given by

$$(2nk)_c = \frac{A_0 \gamma \nu_c}{(\nu_0^2 - \nu_c^2)^2 + \gamma^2 \nu_c^2} \tag{B.1}$$

Correspondingly for a series of classical oscillators resonating at ν_1, ν_2, ν_3, with relative strengths A_1, A_2, A_3 etc. we have an equation analogous to (2.6a), namely

$$(2nk)_c = \frac{A_1 \gamma \nu_c}{(\nu_1^2 - \nu_c^2)^2 + \gamma^2 \nu_c^2} + \frac{A_2 \gamma \nu_c}{(\nu_2^2 - \nu_c^2)^2 + \gamma^2 \nu_c^2} + \frac{A_3 \gamma \nu_c}{(\nu_3^2 - \nu_c^2)^2 + \gamma^2 \nu_c^2} \tag{B.2}$$

In the limit of many oscillators we can replace this expression by an integral with respect to ν, the frequency of the oscillators, replacing the A's by $dA/d\nu$, which gives the distribution of oscillator strength, i.e.

$$(2nk)_c = \int_0^\infty \frac{(dA/d\nu)\gamma \nu_c}{(\nu^2 - \nu_c^2)^2 + \gamma^2 \nu_c^2} \, d\nu \tag{B.3}$$

This integral is evaluated at the limit of $\gamma \to 0$, since we now have effectively an infinite series of oscillators whose individual bandwidths approach zero.

As $\gamma \to 0$ this integral has a significant contribution only around $\nu \doteqdot \nu_c$, so that it is sufficient to integrate between the limits $\nu = \nu_c - r\gamma$ to $\nu = \nu_c + r\gamma$, where $r \gg 1$, but of course $r\gamma \ll \nu_c$. Over this narrow band $dA/d\nu$ can be taken as constant at $dA/d\nu_c$ and we obtain

$$(2nk)_c = (dA/d\nu_c)\gamma \nu_c \int_{\nu_c - r\gamma}^{\nu_c + r\gamma} \frac{d\nu}{(\nu^2 - \nu_c^2)^2 + \gamma^2 \nu_c^2} \tag{B.4}$$

By factorizing the denominator the integral gives

$$(2nk)_c = \frac{(dA/d\nu_c)\gamma \nu_c}{4pq} \left[\log \frac{\nu^2 + p\nu + q}{\nu^2 - p\nu + q} + \frac{2p}{(4q - p^2)^{1/2}} \left(\tan^{-1} \frac{2\nu + p}{\gamma} \right. \right.$$
$$\left. \left. + \tan^{-1} \frac{2\nu - p}{\gamma} \right) \right]_{\nu_c - r\gamma}^{\nu_c + r\gamma} \tag{B.5}$$

where
$$q^2 = v_c{}^4 + \gamma^2 v_c{}^2 \quad \text{and} \quad p^2 = 2q + 2v_c{}^2$$

or
$$q = v_c{}^2 + \tfrac{1}{2}\gamma^2, \quad p = 2v_c + \gamma^2/4v_c, \quad (4q - p^2)^2 = \gamma \quad \text{as} \quad \gamma \to 0.$$

Putting in the limits the logarithm vanishes and the $\tan^{-1}$ terms are each $\tfrac{1}{2}\pi$. Hence

$$(2nk)_c = \frac{\pi}{2v_c} \frac{\mathrm{d}A}{\mathrm{d}v_c} \tag{B.6}$$

which is independent of both r and γ, as it clearly should be.

This relation holds at any value of v_c, so that in general

$$\mathrm{d}A/\mathrm{d}v = (2v/\pi)(2nk) \tag{B.7}$$

The equation for the real part of the dielectric constant may be expressed in terms of a series of classical oscillators, analogous to equation (B.2) and finally in integral form corresponding to equation (B.3),

$$(n^2 - k^2 - 1)_c = \int_0^\infty \frac{(\mathrm{d}A/\mathrm{d}v)(v^2 - v_c{}^2)}{(v^2 - v_c{}^2)^2 + \gamma^2 v_c{}^2} \, \mathrm{d}v \quad \text{as} \quad \gamma \to 0 \tag{B.8}$$

At frequencies not too near to v_c, the γ term in the denominator may be omitted and

$$(n^2 - k^2 - 1)_c = \int_0^{v_c - r\gamma} \frac{(\mathrm{d}A/\mathrm{d}v)}{v^2 - v_c{}^2} \, \mathrm{d}v + \int_{v_c + r\gamma}^\infty \frac{\mathrm{d}A/\mathrm{d}v}{v^2 - v_c{}^2} \, \mathrm{d}v$$
$$+ \int_{v_c - r\gamma}^{v_c + r\gamma} \frac{(\mathrm{d}A/\mathrm{d}v)(v^2 - v_c{}^2) \, \mathrm{d}v}{(v^2 - v_c{}^2)^2 + \gamma^2 v_c{}^2} \tag{B.9}$$

For the third term, we may integrate as before, taking $\mathrm{d}A/\mathrm{d}v = \mathrm{d}A/\mathrm{d}v_c$ as constant. This term gives

$$\left[\frac{(\mathrm{d}A/\mathrm{d}v_c)}{2p} \log \frac{v^2 + pv + q}{v^2 + pv + q} + \frac{\gamma^3}{4} \left(\tan^{-1} \frac{2v + p}{\gamma} + \tan^{-1} \frac{2v - p}{\gamma} \right) \frac{\mathrm{d}A}{\mathrm{d}v_c} \right]_{v_c - r\gamma}^{v_c + r\gamma}$$

This may be seen to become zero as $\gamma \to 0$, irrespective of the value of r. Hence

$$(n^2 - k^2 - 1)_c = \int_0^{v_c - r\gamma} \frac{\mathrm{d}A/\mathrm{d}v}{v^2 - v_c{}^2} \, \mathrm{d}v + \int_{v_c + r\gamma}^\infty \frac{\mathrm{d}A/\mathrm{d}v}{v^2 - v_c{}^2} \, \mathrm{d}v$$
$$= \int_0^\infty \frac{(\mathrm{d}A/\mathrm{d}v)}{v^2 - v_c{}^2} \, \mathrm{d}v \quad \text{as} \quad r \to 0 \tag{B.10}$$

Thus from equation (B.7) we obtain the required inter-relation between the real and imaginary parts of the dielectric constant, namely

$$(n^2 - k^2 - 1)_c = \frac{2}{\pi} \int_0^\infty \frac{2nkv}{v^2 - v_c{}^2} \, \mathrm{d}v \tag{B.11}$$

246

For numerical integration this equation is inconvenient as it stands because the integrand becomes infinite at $\nu = \nu_c$. (This is immaterial in analytical use as the expression is symmetrical about $\nu = \nu_c$.) To remove this difficulty we subtract from (B.11) a second term

$$\frac{2}{\pi} \int_0^\infty \frac{(2nk\nu)_c \, d\nu}{\nu^2 - \nu_c^2}$$

which is readily shown to be zero, and obtain

$$(n^2 - k^2 - 1)_c = \frac{2}{\pi} \int_0^\infty \frac{2nk\nu - (2nk\nu)_c}{\nu^2 - \nu_c^2} \, d\nu \qquad (B.12)$$

This expression, where the integrand is always finite, is the one given already in Chapter 2 (equation 2.26).

B.2 VERIFICATION OF THE INTER-RELATIONS FOR A CLASSICAL ABSORPTION BAND

The above relation should of course hold for a classical oscillator. This may be verified by direct integration as follows:—

Put
$$I_c = \frac{2}{\pi} \int_0^\infty \frac{2nk\nu - (2nk\nu)_c}{\nu^2 - \nu_c^2} \, d\nu$$

where
$$2nk = \frac{A\Gamma\nu}{(\nu_0^2 - \nu^2)^2 + \Gamma^2\nu^2} = \frac{A\Gamma\nu}{(\nu^2 + \alpha\nu + \nu_0^2)(\nu^2 - \alpha\nu + \nu_0^2)} \qquad (B.13)$$

and
$$\alpha^2 = 4\nu_0^2 - \Gamma^2$$

Therefore

$$\frac{\pi}{2} I_c = \frac{A\Gamma/2\alpha}{(\nu_0^2 - \nu_c^2)^2 + \Gamma^2\nu_c^2} \left[\int_0^\infty \frac{\alpha\nu_0^2 + (\nu_0^2 + \nu_c^2)\nu}{\nu^2 + \alpha\nu + \nu_0^2} \, d\nu \right.$$
$$\left. + \int_0^\infty \frac{\alpha\nu_0^2 - (\nu_0^2 + \nu_c^2)\nu}{\nu^2 - \alpha\nu + \nu_0^2} \, d\nu \right]$$
$$+ \frac{A\Gamma}{(\nu_0^2 - \nu_c^2)^2 + \Gamma^2\nu_c^2} \left(\int_0^\infty \frac{\nu_c^2 \, d\nu}{\nu^2 - \nu_c^2} - \int_0^\infty \frac{(2nk\nu)_c \, d\nu}{\nu^2 - \nu_c^2} \right) \qquad (B.14)$$

Now the last two terms are identical and cancel. The first integral is zero, and from the second we obtain

$$\frac{\pi}{2} I_c = \frac{\frac{1}{2}A(\nu_0^2 - \nu_c^2)}{(\nu_0^2 - \nu_c^2)^2 + \Gamma^2\nu_c^2} \left[\tan^{-1}\frac{2\nu + \alpha}{\Gamma} + \tan^{-1}\frac{2\nu - \alpha}{\Gamma} \right]_0^\infty \qquad (B.15)$$

The square bracket is π and hence

$$I_c = \frac{A(\nu_0^2 - \nu_c^2)}{(\nu_0^2 - \nu_c^2)^2 + \Gamma^2\nu_c^2}$$

which is of course the expression for the real part $(n^2 - k^2 - 1)_c$ for the classical oscillator of equation (B.13), at frequency ν_c.

B.3 INTER-RELATION OF n AND k FOR FREE CARRIER ABSORPTION

The formulae for the real and imaginary parts of the dielectric constant are given in equation (2.30). For mathematical convenience we will treat the particular case where $4Ne^2/m\varepsilon_0 = g^2$.

Thus

$$n^2 - k^2 = 1 - g^2/4(\omega^2 + g^2), \qquad 2nk = g^3/4\omega(\omega^2 + g^2) \quad \text{(B.16)}$$

Solving for n and k we obtain

$$n^2 = \frac{\{\omega + (\omega^2 + g^2)^{1/2}\}^3}{8\omega(\omega^2 + g^2)} \quad \text{(B.17)}$$

$$k^2 = \frac{\{(\omega^2 + g^2)^{1/2} - \omega\}^3}{8\omega(\omega^2 + g^2)} \quad \text{(B.18)}$$

Substitute $\omega = g \tan \theta$ and $t^2 = \tan \theta/2$ in (B.18). On simplifying we obtain

$$\int_0^\infty \frac{k\omega \, d\omega}{\omega^2 - a^2} = \int_0^1 \frac{2t^2(1 - t^2)^2 g^2/(1 + t^2) \, dt}{4t^4 g^2 - a^2(1 - t^4)^2} \quad \text{(B.19)}$$

The expression on the right can be factorized to give:

$$\frac{(\alpha + 2)(\beta - \alpha) + \alpha - 2}{4\beta(t^2 - \beta + \alpha)} - \frac{(\alpha + \beta)(\alpha - 2) + \alpha + 2}{4\beta(t^2 - \alpha - \beta)}$$

$$+ \frac{(\alpha + 2)(\alpha + \beta) + 2 - \alpha}{4\beta(t^2 + \alpha + \beta)} - \frac{(\beta - \alpha)(\alpha - 2) - \alpha - 2}{4\beta(t^2 + \beta - \alpha)} - \frac{2}{1 + t^2}$$

where $\alpha = g/a$ and $\beta^2 = 1 + \alpha^2 = 1 + g^2/a^2$.

The integration of the first two terms give logarithms which may be shown to cancel each other. Integration of the last term yields simply $-\pi/2$. The integrals of the third and fourth terms give:—

$$\frac{(\alpha + 2)(\alpha + \beta) + 2 - \alpha}{4\beta(\alpha + \beta)^{1/2}} \tan^{-1} (\alpha + \beta)^{1/2}$$

$$- \frac{(\beta - \alpha)(\alpha - 2) - \alpha - 2}{4\beta(\beta - \alpha)^{1/2}} \tan^{-1}(\beta - \alpha)^{1/2}$$

Since $\beta + \alpha = 1/(\beta - \alpha)$ this becomes

$$\frac{(\beta + \alpha)^{1/2}}{4\beta} \left(\alpha + 2 + \frac{2 - \alpha}{\alpha + \beta} \right) [\tan^{-1}(\alpha + \beta)^{-1/2} + \tan^{-1}(\alpha + \beta)^{1/2}]$$

248

Now the sum of two angles with reciprocal tangents is $\pi/2$. Hence

$$\int_0^\infty \frac{k\omega \, d\omega}{\omega^2 - a^2} = \frac{\pi}{8\beta} (\alpha + \beta)^{1/2} \left(\alpha + 2 + \frac{2 - \alpha}{\alpha + \beta} \right) - \pi/2$$

or

$$\left\{ 1 + \frac{2}{\pi} \int_0^\infty \frac{k\omega \, d\omega}{\omega^2 - a^2} \right\}^2 = \frac{\{a + (g^2 + a^2)^{1/2}\}^3}{8a(g^2 + a^2)}$$

By comparison with equation (B.17) this is seen to be simply $n_a{}^2$. Hence

$$1 + \frac{2}{\pi} \int_0^\infty \frac{k\omega \, d\omega}{\omega^2 - a^2} = n_a$$

which is the desired relation, given in Chapter 2 as equation (2.29).

PHOTOCONDUCTIVITY AT HIGH ILLUMINATION LEVELS

IT has been widely assumed in the past that for high intensity irradiation the photocurrent would increase as the square root of the intensity (see Moss, 1952a).

This result may be obtained theoretically if diffusion is ignored, but the following treatment including diffusion shows that ultimately the photocurrent will increase only as (intensity)$^{1/3}$, although for some conditions of surface recombination a square root law may apply over a restricted range of intensities.

Consider an intrinsic semi-conductor with $n_0 = p_0 = i$, and assume the neutrality condition so that $n = p$ everywhere. Assume surface generation (i.e. highly absorbed radiation) of q electron-hole pairs per unit area per second, and treat a large area relatively thin specimen, so that the problem is essentially one dimensional.

We have

$$\frac{\mathrm{d}p}{\mathrm{d}t} = -r(p^2 - i^2) + D\frac{\mathrm{d}^2p}{\mathrm{d}y^2} \tag{C.1}$$

where the recombination coefficient

$$r = 1/2i\tau \tag{C.2}$$

At equilibrium $\mathrm{d}p/\mathrm{d}t = 0$ and integration of equation (C.1) gives

$$p^3 - 3i^2p = \frac{3D}{2r}\left(\frac{\mathrm{d}p}{\mathrm{d}y}\right)^2 - A \tag{C.3}$$

The current equations are

$$J_y{}^+ = e\mu_p pE_y - eD_p\frac{\mathrm{d}p}{\mathrm{d}y} \tag{C.4}$$

$$J_y{}^- = be\mu_p nE_y + beD_p\frac{\mathrm{d}p}{\mathrm{d}y} \tag{C.5}$$

There is no nett current in the y direction, so

$$J_y{}^+ + J_y{}^- = 0$$

Hence

$$J_y{}^+ = -eD\,\mathrm{d}p/\mathrm{d}y \tag{C.6}$$

where the ambipolar diffusion constant

$$D = \frac{2b}{b+1}D_p \tag{C.7}$$

251

For the boundary condition on the back surface, assume the specimen to be sufficiently thick that:

$$y \to \infty, \quad p \to i \quad \text{and} \quad J_y{}^+ \to 0$$

so that from (C.6)

$$dp/dy \to 0$$

Hence equation (C.3) gives

$$A = 2i^3$$

and

$$p^3 - 3i^2p + 2i^3 = \frac{3D}{2r} \left(\frac{dp}{dy}\right)^2 = 3iL^2 \left(\frac{dp}{dy}\right)^2 \tag{C.8}$$

Integrating gives

$$-y/L = \log \frac{(p + 2i)^{1/2} - (3i)^{1/2}}{(p + 2i)^{1/2} + (3i)^{1/2}} + \log C$$

or

$$(p - i)/3i = 4Ce^{-y/L}(C - e^{-y/L})^{-2} \tag{C.9}$$

Differentiating

$$dp/dy = \frac{-12iC}{L} e^{-y/L} \, (C + e^{-y/L})(C - e^{-y/L})^{-3} \tag{C.10}$$

It will be seen from equations (C.9) and (C.10) that the boundary conditions assumed for the back surface will be met satisfactorily if the specimen thickness is a few diffusion lengths. For the front surface the boundary conditions are

$$y = 0, \qquad J_y{}^+ = eq - es(p - i) = -eD \, dp/dy \tag{C.11}$$

giving

$$q - 12iCs(C - 1)^{-2} = 12iDL^{-1} C(C + 1)(C - 1)^{-3}$$

or

$$q\tau = 12iLC(C + 1 + \alpha C - \alpha)(C - 1)^{-3} \tag{C.12}$$

where

$$\alpha = \tau s/L = s(\tau/D)^{1/2}$$

The total excess holes per unit area is given by

$$P = \int_0^\infty (p - i) \, dy = -3i \left[\frac{4CL}{C - e^{-y/L}}\right]_0^\infty = 12iL/(C - 1) \tag{C.13}$$

Hence from equations (C.12) and (C.13),

$$q\tau = P(1 + P/12iL)(1 + \alpha + 2P/12iL) \tag{C.14}$$

The photocurrent for small fields applied in either the x or z directions will be directly proportional to P, so that equation (C.14) gives the form of the signal-intensity law. For small intensities, where $P \ll 12iL$ the relation, as expected becomes

$$P = q\tau \tag{C.15}$$

For high intensities where $P \gg 12iL(1 + \alpha)$

$$P^3/72i^2L^2 = q\tau$$

or

$$P \propto q^{1/3} \qquad (C.16)$$

i.e. a cube root law is obtained.

If $\alpha \gg 1$, then it is possible theoretically to have an intermediate region where $P/12iL \gg 1$ but $\ll \frac{1}{2}(1 + \alpha)$, and in this region (C.14) will approximate to $P \propto q^{1/2}$. For the higher mobility semi-conductors this range will be very restricted, but for materials with low mobilities and high τ values the range could be considerable. Ultimately however the cube root law would be expected.

Equation (C.14) is plotted in *Figure C.1* for two values of α, the low value

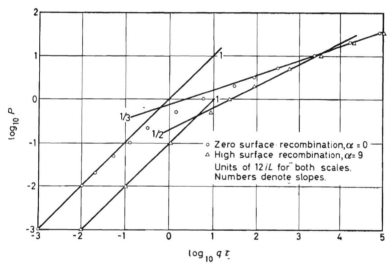

Figure C.1. Theoretical photocurrent-intensity relation

of α showing the cube law behaviour while the high value of α gives a significant range of square root dependence.

If diffusion is ignored, i.e. D is tacitly equated to zero, then α becomes infinite and a square root law will be obtained theoretically for $P/12iL \gg 1$. Diffusion can clearly be ignored in the simple case of uniform carrier generation throughout the bulk of the specimen and no surface recombination. This state of affairs tends to occur if the specimen is thin compared with the reciprocal of the absorption constant, so that it might well happen in thin films or on irradiation with wavelengths near the absorption edge. Under these conditions therefore, a square root law at high intensities is quite likely.

REFERENCE LIST AND AUTHOR INDEX

Italic figures indicate page numbers

ABRAHAM, A. *See* TAUC, J.

AIGRAIN, P. (1954a) *Ann. Rad. Comp. Gen. T.S.F.* **15** 52: *63, 94*

— (1954b) *Physica* **20** 1010: *91*

— and BULLIARD H. (1953) *C.R. Acad. Sci., Paris* **236** 595, 672: *63*

— and DES CLOIZEAUX, J. (1955) *C.R. Acad. Sci., Paris* **241** 859: *46*

— and GARRETA, O. (1954) *C.R. Acad. Sci., Paris* **238** 1573: *76*

AKA, E. Z. *See* STRAUMARIS, M. E.

ALEXANDER, B. H. *See* LEVITAS, A.

ALINE, P. G. (1957) *Phys. Rev.* **105** 406: *39*

ALLEMAND, C. and ROSSEL, J. (1954) *Helv. phys. acta* **27** 519: *107*

ALLEN, J. W. *See* MACKINTOSH, I. M.

ALLEN, R. C. *See* CZYZAK, S. J.; REYNOLDS, D. C.

ANTONCIK, E. (1955) *Czech. J. Phys.* **5** 449: *137*

— (1956) *Czech. J. Phys.* **6** 209: *116*

APFEL, J. H. and HADLEY, L. N. (1955) *Phys. Rev.* **100** 1689: *48*

APKER, L. *See* HUNTINGTON, H. B.; TAFT, E.

— and TAFT, E. (1949) *Phys. Rev.* **75** 1181: *53*

— — (1954) *Phys. Rev.* **96** 1496: *53*

— — and DICKEY, J. (1948) *Phys. Rev.* **74** 1462: *53*

APPEL, J. (1954) *Z. Naturf.* **9a** 265: *202*

— and LAUTZ, G. (1954) *Physica* **20** 1110: *202, 221*

ARCHER, R. J. (1958) *Phys. Rev.* **110** 354: *136*

ARON, J. and GROETZINGER, G. (1955) *Phys. Rev.* **100** 1128: *63*

ARSENEVA-GAIL, A. N. (1949) *C. R. Acad. Sci. U.R.S.S.* **68** 245: *53*

— (1955) *J. tech. Phys., Moscow* **25** 1544: *53*

ARTHUR, J. B., GIBSON, A. F. and GRANVILLE, J. W. (1956) *J. Electronics* **2** 145: *140*

AUSTIN, I. G. and McCLYMONT, D. R. (1954) *Physica* **20** 1077: *230*

— and WOLFE, R. (1956) *Proc. phys. Soc., Lond.* **B69** 329: *107, 111*

AVERBACH, B. L. *See* KOLM, C.

AVERY, D. G. (1951) *Proc. phys. Soc., Lond.* **B64** 1087: *24*

— (1952) *Proc. phys. Soc., Lond.* **B65** 425: *9, 187*

— (1953) *Proc. phys. Soc., Lond.* **B66** 133: *9, 24, 187, 191*

— (1954) *Proc. phys. Soc., Lond.* **B67** 2: *28, 187, 190, 191, 198*

— and CLEGG, P. L. (1953) *Proc. phys. Soc., Lond.* **B66** 512: *135*

— GOODWIN, D. W., LAWSON, W. D. and MOSS, T. S. (1954) *Proc. phys. Soc., Lond.* **B67** 761: *41*

— — and RENNIE, A. E. (1957) *J. sci. Instrum.* **34** 394: *240*

BADER, L. J. and JACOBSMEYER, V. P. (1954) *Bull. Amer. phys. Soc.* **29** 12: *102*

BAKER, W. M. *See* CZYZAK, S. J.

BALKANSKI, M. (1958) *Rochester Semiconductor Conference*; to be published in *J. Phys. Chem. Solids*, 1959: *220*

BARDEEN, J. (1949) *Phys. Rev.* **75** 1777: *169*

— BLATT, F. J. and HALL, L. H. (1954/6) *Photoconductivity Conference*; Wiley, N.Y.: *36, 37*

— and SHOCKLEY, W. (1950) *Phys. Rev.* **80** 72: *44, 45, 118, 170*

BARNARD, G. P. (1930) *The Selenium Cell*, Constable,: London *171*

BARNES, R. B., BRATTAIN, R. R. and SEITZ, F. (1935) *Phys. Rev.* **48** 582: *28, 52*
BARRIE, R. *See* HILSUM, C.
BARTSCHAT, A. *See* GOBRECHT, H.
— and GOBRECHT, H. (1958) *Halbleiter und Phosphore* (Garmisch Semiconductor Conference); Vieweg, Braunschweig: *216*
BASTIN, J. A. *See* WRIGHT, R. W.
BATE, G. *See* STARKIEWICZ, J.
BECKER, J. H. *See* BRECKENRIDGE, R. G.
— (1955) *Phys. Rev.* **98** 1192: *179*
BECKER, M. *See* FAN, H. Y.
BECQUEREL, F. (1877) *Ann. Chim. (Phys.)* **12** 1: *156, 212*
BEDO, D. E. *See* TOMBOULIAN, D. H.
BEER, A. C. *See* HARMAN, T. C.; WILLARDSON, R. W.
BELL, D. G., HUM, D. M., PINCHERLE, L., SCIAMA, D. W. and WOODWARD, P. M. (1953) *Proc. Roy. Soc.* **A217** 71: *182, 190*
BELL, E. E. *See* BURSTEIN, E.
BEMSKI, G. *See* BITTMAN, C. A.
BENEDICT, T. S. (1953) *Phys. Rev.* **91** 1565: *132*
BENNET, H. *See* STARKIEWICZ, J.
BENNETT, H. E. *See* RANK, D. H.
BENOIT À LA GUILLAUME, C. (1956) *C. R. Acad. Sci., Paris* **243** 704: *145*
BERCKHEMER, H. *See* RICHTER, H.
BERGMAN, L. and HANSLER, J. (1936) *Z. Phys.* **100** 50: *206*
BERLAGA, R. YA. and STRAKHOV, L. P. (1954) *J. tech. Phys., Moscow* **24** 943: *200*
BERNARD, M. (1958) *Rochester Semiconductor Conference*; to be published in *J. Phys. Chem. Solids*, 1959: *91*
BERRY, R. E. *See* SOMMERS, H. S.
BESS, L. (1956) *Phys. Rev.* **103** 72: *200*
— (1957) *Phys. Rev.* **105** 1469: *98*
BILENKO, D. I. *See* KYRYASHKINA, E. I.
BILLIG, E. (1952) *Proc. phys. Soc., Lond.* **B65** 216: *163*
BIRMAN, J. *See* SHAKIN, C.
BITTMAN, C. A. and BEMSKI, G. (1957) *J. appl. Phys.* **28** 1423: *123*
BLAKEMORE, J. S. (1956) *Canad. J. Phys.* **34** 938: *127*
BLATT, F. J. *See* BARDEEN, J.; BURSTEIN, E.
BLET, G. (1956) *C.R. Acad. Sci., Paris* **242** 95: *161, 164*
BLOEM, J. (1956a) *Philips Res. Rep.* **11** 273: *181, 186*
— (1956b) *Appl. sci. Res., Hague* **B6** 92: *200*
— and KRÖGER, F. A. (1956) *Z. phys. Chem.* **7** 1: *181*
BLOUNT, E., CALLAWAY, J., COHEN, M., DUMKE, W. P. and PHILLIPS, J. (1956) *Phys. Rev.* **101** 563: *228*
BLUNT, R. F. *See* BRECKENRIDGE, R. G.; FREDERIKSE, H. P. R.
BODE, H. W. (1945) *Network Analysis and Feedback Amplifier Design*, Van Nostrand, London and New York: *24*
BOND, W. L. *See* MCSKIMIN, H. J.
BORISSOV, M. and KANEV, ST. (1955) *Z. phys. Chem. Lpz.* **205** 56: *206*
BOTTOM, V. E. (1948) *Phys. Rev.* **74** 1218: *173*
BOYLE, W. S. and BRAILSFORD, A. D. (1957) *Phys. Rev.* **107** 903: *228, 229, 234*
BRAILSFORD, A. D. *See* BOYLE, W. S.
BRAITHWAITE, J. G. N. (1951) *Proc. phys. Soc., Lond.* **B64** 274: *206*
— (1955) *J. sci. Instrum.* **32** 10: *185, 198*
BRATTAIN, R. R. *See* BARNES, R. B.

BRATTAIN, W. *See* BUCK, T. M.
— (1951) *Semiconductor Materials* (Reading conference); Butterworths, London: *63*
BRAUNSTEIN, R. (1955) *Phys. Rev.* **99** 1892: *91, 225*
— MOORE, A. R. and HERMAN, F. (1958) *Phys. Rev.* **109** 695: *150, 151*
BREBRICK, R. F. *See* SCANLON, W. W.
— and SCANLON, W. W. (1957) *J. chem. Phys.* **27** 607: *193*
BRECKENRIDGE, R. G., BLUNT, R. F., HOSLER, W. R., FREDERIKSE, H. P. R.,
    BECKER, J. H. and OSHINSKY, W. (1954) *Phys. Rev.* **96** 571: *232*
BREITLING, G. *See* RICHTER, H.
BRIGGS, H. B. *See* HAYNES, J. R.; TANENBAUM, M.
— (1950) *Phys. Rev.* **77** 287: *17, 115, 120, 131*
— CUMMINGS, R. F., HROSTOWSKI, H. J. and TANENBAUM, M. (1954) *Phys. Rev.*
    93 912: *48*
— and FLETCHER, R. C. (1952) *Phys. Rev.* **87** 1130: *51*
— — (1953) *Phys. Rev.* **91** 1342: *139*
BRIGMAN, P. W. (1938) *Proc. Amer. Acad. Arts Sci.* **72** 157: *169*
BRODER, J. D. *See* WOLFE, G. A.
BROOKS, H. *See* CARDONA, M.; WARSCHAUER, D. M.
— (1955) *Advances in Electronics and Electron Physics* Vol. 7; Academic Press, N.Y.:
    31, 34, 36, 115, 119, 138, 149
BROPHY, J. J. (1955) *Phys. Rev.* **99** 1336: *111*
— (1956) *Physica* **22** 156: *107, 112*
BROPHY, V. A. *See* LEMPICI, A.
BROSER, I. (1957) *Brooklyn Polytechnic Symposium* **7** 313: Interscience, N.Y.: *218*
— and BROSER–WARMINSKY, R. (1955) *Brit. J. appl. Phys. Sup.* **4** 90: *214*
BROUDE, V. L., EREMENKO, V. V. and RASHBA, E. I. (1957) *Dokl. Akad. Nauk.*
    U.S.S.R. **114** 520: *216*
BROWN, S. C. *See* GOLDEY, J. M.
BUBE, R. H. *See* JENNY, D. A.
— (1953) *J. chem. Phys.* **21** 1409: *206, 218, 219*
— (1955a) *Proc. Inst. Radio Engrs, N.Y.* **43** 1836: *201, 202, 204, 220*
— (1955b) *Phys. Rev.* **98** 431: *206, 221*
— (1956) *Photoconductivity Conference*; Wiley, N.Y.: *206*
— (1957) *J. Phys. Chem. Solids* **1** 234: *220*
— and LIND, E. L. (1958) *Phys. Rev.* **110** 1040: *206*
BUCK, T. M. and BRATTAIN, W. (1955) *Trans. Electrochem. Soc.* **102** 636: *63*
— and MCKIM, F. S. (1957) *Phys. Rev.* **106** 904: *143*
BULLIARD, H. *See* AIGRAIN, P.
— (1954a) *Ann. Phys. Paris* **9** 52: *63*
— (1954b) *Phys. Rev.* **94** 1564: *63, 67*
BURSTEIN, E. *See* LAX, M.; PICUS, G. S.; SCLAR, N.
— (1954) *Phys. Rev.* **93** 632: *45, 231*
— and DAVISSON, J., BELL, E. E., TURNER, W. J. and LIPSON, H. G. (1954) *Phys.
    Rev.* **93** 65: *148*
— and EGLI, P. H. (1955) *Advances in Electronics and Electronic Physics* Vol. 7;
    Academic Press, N.Y.: *115, 211, 235*
— and PICUS, G. S. (1957) *Phys. Rev.* **105** 1123: *90, 228*
— — (1958) *Brussels Solid State Congress*; Academic Press, N.Y.: *128*
— — and GEBBIE, H. A. (1956a) *Phys. Rev.* **103** 825 826: *46, 47, 88, 229, 233, 234*
— — HENVIS, B. W. and LAX, M. (1955) *Bull. Amer. phys. Soc.* **30** 13: *124, 126*
— — and SCLAR, N. (1956b) *Photoconductivity Conference*; Wiley, N.Y.: *88, 113, 121,*
    122, 124, 125, 131, 139, 140, 145

257

BURSTEIN, E., PICUS, G. S., WALLIS, R. F. and BLATT, F. J. (1958/59) *Rochester Semiconductor Conference*; to be published in *J. Phys. Chem. Solids*, 1959: also *Phys. Rev.* in press: *90*

BUSCH, G. (1958) *Halbleiter und Phosphore* (Garmisch Semiconductor Conference); Vieweg, Braunschweig: *179*

— and MOOSER, E. (1953) *Helv. phys. acta* **26** 611: *179*

— and SCHNEIDER, M. (1954) *Physica* **20** 1084: *97*

BYKOVSKII, YU. A. *See* KIKOIN, I. K.

CALLAWAY, J. *See* BLOUNT, E.

CALLEN, H. P. (1954) *J. chem. Phys.* **22** 518: *173*

CANNON, C. G. *See* GEBBIE, H. A.

CARDONA, M., PAUL, W. and BROOKS, H. (1958) *Brussels Solid State Congress*; Academic Press, N.Y.: also *Rochester Semiconductor Conference*; to be published in *J. Phys. Chem. Solids*, 1959: *116, 133*

CARLSON, R. O. *See* COLLINS, C. B.

— (1957) *Phys. Rev.* **108** 1390: *127*

CARVALHO, A. P. DE. *See* DE CARVALHO, A. P.

CASHMAN, R. J. *See* CLARK, M. A.

CASPARI, M. E. *See* RAU, R. R.

CHABAN, M. M. *See* KLINGER, M. I.

CHAMPION, F. C. and DALE, B. (1956) *Proc. Roy. Soc.* **A234** 419: *107*

— and HUMPHREYS, D. L. O. (1957) *Proc. phys. Soc., Lond.* **B70** 320: *107, 111*

CHAPIN, D. M., FULLER, C. S. and PEARSON, G. L. (1954) *J. appl. Phys.* **25** 676: *82, 122*

CHASMAR, R. P. *See* SMITH, R. A.

— (1956) *Photoconductivity Conference*; Wiley, N.Y.: *181*

— and STRATTON, R. (1956) *Phys. Rev.* **102** 1686: *46*

CHECINSKA, H. and SOSNOWSKI, L. (1954) *Bull. Acad. Sci., Polon.* Cl 3 **2** 383: *195*

CHOLET, P. *See* LASSER, M. E.

CHOYKE, W. J. and PATRICK, L. (1957a) *Phys. Rev.* **105** 1721: *112*

—— (1957b) *Phys. Rev.* **108** 25: *161*

CHRISTIAN, S. M. *See* JOHNSON, E. R.

CHYNOWETH, A. G. and McKAY, K. G. (1956) *Phys. Rev.* **102** 369: *91*

CLARK, C. D. (1958) *Malvern Conference on Spectroscopy of Solids*; *Nature, Lond.* **182** 159: also *Rochester Semiconductor Conference*; to be published in *J. Phys. Chem. Solids*, 1959: *111*

— DITCHBURN, R. W. and DYER, H. B. (1956) *Proc. Roy. Soc.* **A234** 363: *111*

CLARK, M. A. and CASHMAN, R. J. (1952) *Phys. Rev.* **85** 1043: *183*

CLEGG, P. L. *See* AVERY, D. G.

CLOIZEAUX, J. DES. *See* DES CLOIZEAUX, J.

COHEN, J. (1954) *J. appl. Phys.* **25** 798: *179*

COHEN, M. *See* BLOUNT, E.

COLLINS, C. B. and CARLSON, R. O. (1957) *Phys. Rev.* **108** 1409: *123, 127*

—— and GALLAGHER, C. J. (1957) *Phys. Rev.* **105** 1168: *124*

COLLINS, R. J. *See* FAN, H. Y.; KAISER, W.

— and FAN, H. Y. (1954) *Phys. Rev.* **93** 674: *112, 121, 141*

COOGAN, C. K. (1957) *Proc. phys. Soc., Lond.* **70** 845: *207, 208, 209, 210*

COPE, A. D. *See* WEIMER, P. K.

CRANCE, R. C. *See* CZYZAK, S. J.

CRESSELL, I. G. and POWELL, J. A. (1957) *Progress in Semiconductors*, Vol. 2 137; Heywood, London: *129, 143*

CRONEMEYER, D. C. *See* RANK, D. H.
— (1957) *Phys. Rev.* **105** 522: *113, 114*
CUMMERLOW, R. L. (1954) *Phys. Rev.* **95** 16, 561: *82, 122*
CUMMINGS, R. F. *See* BRIGGS, H. B.
CUNNELL, F. A. and SAKER, E. W. (1957) *Progress in Semiconductors*, Vol. 2 37;
    Heywood, London: *48, 223*
CUSTERS, J. F. H. (1954) *Physica* **20** 183: *107*
— (1955) *Nature, Lond.* **176** 173: *111*
— and NEAL, F. A. (1957) *Nature, Lond.* **179** 268: *111*
CZYZAK, S. J. *See* REYNOLDS, D. C.
— BAKER, W. M., CRANCE, R. C. and HOWE, J. B. (1957) *J. opt. Soc. Amer.* **47**
    240: *209, 210, 217*
— RENOLDS, D. C., ALLEN, R. C. and REYNOLDS, C. C. (1954) *J. opt. Soc. Amer.*
    44 864: *207*

DALE, B. *See* CHAMPION, F. C.
DANIELSON, G. C. *See* SHAW, W. C.
DASH, W. C. (1955) *Phys. Rev.* **98** 1536: *116*
— and NEWMAN, R. (1955) *Phys. Rev.* **99** 1151: *116, 117, 118, 133, 134, 137*
DAVIS, W. C. *See* SINTON, W. M.
DAVIS, W. D. (1958) *J. appl. Phys.* **29** 231: *128*
DAVISSON, J. *See* BURSTEIN, E.
DE CARVALHO, A. P. (1956) *C.R. Acad. Sci., Paris* **242** 745: *73*
— (1957) *C.R. Acad. Sci., Paris* **244** 461: *178*
DECHENE, G. (1939) *C.R. Acad. Sci., Paris* **208** 95: *206*
DEHLINGER, U. (1957) *J. Phys. Chem. Solids* **1** 279: *107*
DELL, R. M. *See* REYNOLDS, W. N.
DEMBER, H. (1931) *Phys. Z.* **32** 554: *61*
DE NOBEL, D. *See* KRÖGER, F. A.; VAN DOORN, C. Z.
— and HOFMAN, D. (1956) *Physica* **22** 252: *221*
DES CLOIZEAUX, J. *See* AIGRAIN, P.
DESTRIAU, G. and IVEY, H. F. (1955) *Proc. Inst. Radio Engrs, N.Y.* **43** 1911: *98*
DE VAUX, L. H. and PIZZARELLO, F. A. (1956) *Phys. Rev.* **102** 85: *230*
DEXTER, D. L. (1954/6) *Photoconductivity Conference*; Wiley, N.Y.: *38, 47*
— (1956) *Phys. Rev.* **101** 48: *36, 38*
— (1958) *Rochester Semiconductor Conference*; to be published in *J. Phys. Chem. Solids*,
    1959: *39, 216*
DEXTER, R. N. *See* LAX, B.; ZEIGER, H. J.
— (1958) *Rochester Semiconductor Conference*; to be published in *J. Phys. Chem. Solids*,
    1959: *217*
— LAX, B., KIP, A. F. and DRESSELHAUS, G. (1954a) *Phys. Rev.* **96** 222: *114*
— ZEIGER, H. J. and LAX, B. (1954b) *Phys. Rev.* **95** 557: *129*
— — — (1956) *Phys. Rev.* **104** 637: *88*
DICKEY, J. *See* APKER, L.
DIEMER, G. and HOOGENSTRAATEN, W. (1957) *J. Phys. Chem. Solids* **2** 119: *47, 220*
DINGLE, R. B. (1952) *Proc. Roy. Sqc.* **A212** 38: *88*
— (1955) *Phys. Rev.* **99** 1901: *31*
DITCHBURN, R. W. *See* CLARK, C. D.
— (1952) *Light*; Blackie, London: *16*
— (1955) *J. opt. Soc. Amer.* **45** 743: *9*
DIXON, J. R. (1957) *Phys. Rev.* **107** 374: *225*
DOUGLAS, R. W. *See* GOODMAN, C. H. L.

Dowd, J. J. (1951) *Proc. phys. Soc., Lond.* **B64** 783: *153, 154, 157, 161*
Drahokoupil, J., Malkovska, M. and Tauc, J. (1957) *Czech. J. Phys.* **7** 57: *142*
Dresselhaus, G. *See* Dexter, R. N.
— (1957) *Phys. Rev.* **105** 135: *167, 168, 174, 175*
— Kip, A. F. and Kittel, C. (1955a) *Phys. Rev.* **98** 368; **100** 618: *89, 114*
— — — and Wagoner, G. (1955b) *Phys. Rev.* **98** 556: *227, 234*
— Wagoner, G., Kip, A. F. and Han Ying Ku (1955) *Phys. Rev.* **100** 1218: *150*
Drude, P. (1900) *Phys. Z.* **1** 161: *30*
— (1902) *Theory of Optics*; Longmans, London: *243*
Dubrovski, G. B. *See* Maslakovets, Yu P.
Dumbleton, M. J. *See* Garlick, G. F. J.
Dumke, W. P. *See* Blount, E.
— (1957a) *Phys. Rev.* **105** 139: *96, 97, 98*
— (1957b) *Phys. Rev.* **108** 1419: *227, 228*
Dunlap, W. C. (1957) *Introduction to Semiconductors*; Wiley, N.Y.: *114, 129, 130, 145, 218*
— and Watters, R. L. (1953) *Phys. Rev.* **92** 1396: *116, 132*
Dunstadter, H. (1954) *Z. Phys.* **137** 383: *203*
Dutton, D. (1956) *Photoconductivity Conference*; Wiley, N.Y.: *200*
Dyer, H. B. *See* Clark, C. D.
— and Matthews, I. G. (1958) *Proc. Roy. Soc.* **A243** 320: *112*
— and Wedepohl, P. T. (1956) *Proc. phys. Soc., Lond.* **B69** 410: *111*

Eckart, F. and Rabenhorst, H. (1957) *Ann. Phys., Lpz.* **19** 381: *158*
Egli, P. H. *See* Burstein, E.
Ehrenreich, H. (1957) *J. Phys. Chem. Solids* **2** 131: *231, 235*
Elliot, R. J. (1957) *Phys. Rev.* **108** 1384: *48*
Enright, D. P. *See* Talley, R. M.
Eremenko, V. V. *See* Broude, V. L.
Ewald, A. W. *See* Goland, A. N.; Kohnke, E. E.
— and Kohnke, E. E. (1955) *Phys. Rev.* **97** 607: *179*
— and Tufte, O. N. (1958) *J. appl. Phys.* **29** 1007: *179*

Fan, H. Y. *See* Collins, R. J.; Kaiser, W.; Spitzer, W. G.
— (1951) *Phys. Rev.* **82** 900: *45*
— (1955) *Solid State Physics* **1** 283; Academic Press, N.Y.: *129*
— (1956) *Rep. Progr. Phys.* **19** 107: *31, 35, 36, 45, 50, 113, 120, 145, 146, 171, 172, 173, 174*
— and Becker, M. (1951) *Semiconductor Materials* (Reading Conference); Butterworths, London: *32, 120, 138*
— Fisher, P. and Ramdas, A. K. (1958) *Rochester Semiconductor Conference*; to be published in *J. Phys. Chem. Solids*, 1959: *145*
— Kaiser, W., Klontz, E. E., Lark-Horowitz, K. and Pepper, R. R. (1954) *Phys. Rev.* **95** 1087: *40*
— Shepherd, M. L. and Spitzer, W. (1954/56) *Photoconductivity Conference*; Wiley, N.Y.: *37, 51, 113, 116, 117, 118, 119, 137*
→ — Spitzer, W. and Collins, R. J. (1956) *Phys. Rev.* **101** 566: *138*
Fedorus, G. A. *See* Lashkarev, V. E.
Feild, N. J. *See* Wolfe, G. A.
Feltynowski, A., Glass, I., Privkowski, T. and Torun, T. (1954) *Bull. Acad. Sci. Polon.* C13 **2** 389: *195*
Ferguson, W. F. C. *See* Gottesman, J.; Hall, J. F.
Filinski, I. (1957) *Phys. Rev.* **107** 1193: *141*

260

FINE, S. *See* RITTNER, E. S.

FISCHER, F., GUDDEN, B. and TREU, M. (1938) *Phys. Z.* **39** 127: *194*

FISCHER, G., WHITE, G. K. and WOODS, S. B. (1957) *Phys. Rev.* **106** 480: *167*

FISHER, P. *See* FAN, H. Y.

FLETCHER, R. C. *See* BRIGGS, H. B.

FLOOD, W. F. *See* HROSTOWSKI, H. J.

FOCHS, P. D. (1956) *Proc. phys. Soc., Lond.* **B69** 70: *42, 154, 162*

FONER, S. *See* KEYES, R. J.; ZWERDLING, S.

FOURIE, D. J. (1954) *Naturwissenschaften* **41** 328: *98*

FOX, J. J. *See* ROBERTSON, R.

FRANKL, D. R. *See* LEMPICI, A.

FRANKS, J. (1957) *Proc. phys. Soc., Lond.* **B70** 892: *180*

FREDERIKSE, H. P. R. *See* BRECKENRIDGE, R. G.

— and BLUNT, R. F. (1956) *Photoconductivity Conference*; Wiley, N.Y.: *61*

— and HOSLER, W. R. (1957) *Phys. Rev.* **108** 1136, 1146: *227*

FRENKEL, J. (1934) *Phys. Z. Sowjet.* **5** 597: *67*

— (1935) *Phys. Z. Sowjet.* **8** 135: *67*

FRERICHS, R. (1947) *Phys. Rev.* **72** 594; (also unpublished wartime work): *214, 215, 218, 221*

FREYMANN, R. and STIEBER, A. (1934) *C.R. Acad. Sci., Paris* **199** 1109: *105*

FUJIOKA, Y. (1932) *Z. Phys.* **76** 537: *29*

FUKUROI, T. (1949) *Sci. Rep. Res. Insts. Tôhoku Univ.* **A1** 365, 375: *167*

— (1950) *Sci. Rep. Res. Insts. Tôhoku Univ.* **A2** 233, 239: *167*

— (1951) *Sci. Rep. Res. Insts. Tôhoku Univ.* **A3** 175: *170*

— and YAMANOUCHI, C. (1957) *Sci. Rep. Res. Insts. Tôhoku Univ.* **A9** 262: *227*

FULLER, C. S. *See* CHAPIN, D. M.

GALKIN, L. N. and KOROLEV, N. V. (1953) *Dokl. Akad. Nauk U.S.S.R.* **92** 529: *193*

GALLAGHER, C. J. *See* COLLINS, C. B.

GARLICK, G. F. J. *See* HALPERIN, A.

— (1956) *Handbuch der Physik* **19** 316; Springer-Verlag, Berlin: *201, 203*

— and DUMBLETON, M. J. (1954) *Proc. phys. Soc., Lond.* **B67** 442: *193*

GARRETA, O. *See* AIGRAIN, P.

— and GROSVALET, J. (1956) *Progress in Semiconductors* Vol. 1 165; Heywood, London: *63, 76*

GARTNER, W. (1957) *Phys. Rev.* **105** 823: *63, 71*

GEBALLE, T. H. *See* HROSTOWSKI, H. J.

GEBBIE, H. A. *See* BURSTEIN, E.

— and CANNON, C. G. (1952) *J. opt. Soc. Amer.* **42** 277: *156*

— and KIELY, D. G. (1952) *Proc. phys. Soc., Lond.* **B65** 553: *157, 158*

— and SAKER, E. W. (1951) *Proc. phys. Soc., Lond.* **B64** 360: *154, 156, 157*

GELLER, S. *See* HOARD, J. L.

GERRITSEN, H. J. *See* RUPPEL, W.

GIBSON, A. F. *See* ARTHUR, J. B.

— (1950) *Proc. phys. Soc., Lond.* **B63** 756: *28, 185, 190*

— (1952a) *Proc. phys. Soc., Lond.* **B65** 378: *28, 183, 184, 185, 186, 190, 192, 194, 198*

— (1952b) *Proc. phys. Soc., Lond.* **B65** 196: *190, 194, 195*

— (1953) *Proc. phys. Soc., Lond.* **B66** 588: *29, 140*

— (1956) *Proc. phys. Soc., Lond.* **B69** 488: *30*

— and GRANVILLE, J. W. (1957) *Brooklyn Polytechnic Symposia* **7** 303; Interscience, N.Y.: *29, 140*

GILLEO, M. A. (1951) *J. chem. Phys.* **19** 1291: *152, 153, 154, 158, 159, 160, 164*

GIVENS, M. P. *See* WOODRUFF, R. W.
— KOESTER, C. J. and GOFFE, W. I. (1955) *Phys. Rev.* **100** 1112: *161, 174*
— and SIEGMUND, W. P. (1952) *Phys. Rev.* **85** 313: *43*
GLASS, I. *See* FELTYNOWSKI, A.
GLICKSMAN, M. *See* HERMAN, F.
GOBRECHT, H. *See* BARTSCHAT, A.
— and BARTSCHAT, A. (1953) *Z. Phys.* **136** 224: *215, 219*
— and HAMISCH, H. (1957) *Z. Phys.* **148** 218: *163*
GODEFROY, L. R. (1956) *Progress in Semiconductors* Vol. 1 195; Heywood, London: *173*
GOFFE, W. I. *See* GIVENS, M. P.
GOLAND, A. N. and EWALD, A. W. (1956) *Phys. Rev.* **104** 948: *179*
GOLDBERG, A. E. *See* MITCHELL, G. R.
— and MITCHELL, G. R. (1954) *J. chem. Phys.* **22** 220: *194*
GOLDEY, J. M. and BROWN, S. C. (1955) *Phys. Rev.* **98** 1761: *132*
GOLDSTEIN, B. (1958) *Phys. Rev.* **109** 601: *222*
GOODMAN, C. H. L. (1955) *J. Electronics* **1** 115: *48*
— and DOUGLAS, R. W. (1954) *Physica* **20** 1107: *202*
GOODWIN, D. W. *See* AVERY, D. G.
— (1956) *Semiconductor Conference (Rugby) Rep.*; Phys. Soc. (London): *226*
— (1957) *J. sci. Instrum.* **34** 367: *241*
GOTTESMAN, J. and FERGUSON, W. F. C. (1954) *J. opt. Soc. Amer.* **44** 368: *214, 217, 218*
GRANVILLE, J. W. *See* ARTHUR, J. B.; GIBSON, A. F.
GREINER, E. S. and GUTOWSKI, J. A. (1957) *J. appl. Phys.* **28** 1364: *102*
GRISON, E. (1951) *J. chem. Phys.* **19** 1109: *170*
GROETZINGER, G. *See* ARON, J.
GROSS, E. F., KAPLYANSKII, A. A. and NOVIKOV, B. V. (1956) *J. tech. Phys., Moscow* **26** 913: *220*
— and KARRYEV, N. A. (1952) *Dokl. Akad. Nauk U.S.S.R.* **84** 261 471: *48*
— and SOBOLEV, V. V. (1956) *J. tech. Phys., Moscow* **26** 1622: *204*
— and YAKOBSON, M. A. (1955) *Dokl. Akad. Nauk U.S.S.R.* **102** 485: *216*
— — (1956) *J. tech. Phys., Moscow* **26** 6 1369: *216*
GROSVALET, J. *See* GARRETO, O.
— (1954) *Ann. Rad. Comp. Gen. T.S.F.* **9** 360: *63*
GUDDEN, B. *See* FISCHER, F.
GURNEY, R. W. *See* MOTT, N. F.
GUTOWSKY, J. A. *See* GREINER, E. S.

HADLEY, L. N. *See* APFEL, J. H.
HALL, J. F. (1955) *Phys. Rev.* **97** 1471: *67*
— (1956) *J. opt. Soc. Amer.* **46** 1013: *207, 208, 210, 214, 217, 218*
— and FERGUSON, W. F. C. (1955) *J. opt. Soc. Amer.* **45** 714: *209, 210, 216, 217, 218*
HALL, L. H. *See* BARDEEN, J.
HALPERIN, A. and GARLICK, G. F. J. (1955) *Proc. phys. Soc., Lond.* **B68** 758: *216*
HALSTEAD, R. E. (1957) *J. appl. Phys.* **28** 1131: *122*
HAMILTON, D. R. (1958) *Brit. J. appl. Phys.* **9** 103: *201*
HAMISCH, H. *See* GOBRECHT, H.
HANSLER, J. *See* BERGMAN, L.
HAN YING KU. *See* DRESSELHAUS, G.
HARADA, R. H. and MINDEN, H. T. (1956) *Phys. Rev.* **102** 1258: *195*

REFERENCE LIST AND AUTHOR INDEX

HARMAN, T. C. *See* WILLARDSON, R. K.

— WILLARDSON, R. K. and BEER, A. C. (1955) *Phys. Rev.* **98** 1532: *226*

HARRICK, N. J. (1956) *Phys. Rev.* **101** 491; **103** 1173: *30, 140*

HARRISON, S. E. (1954) *Phys. Rev.* **93** 52: *202*

HARTIG, P. A. and LOFERSKI, J. J. (1954) *J. opt. Soc. Amer.* **44** 17: *170, 171*

HAWKINS, T. D. H. *See* MOSS, T. S.

HAYNES, J. R. (1955) *Phys. Rev.* **98** 1866: *91, 143, 144*

— (1958) *Halbleiter und Phosphore* (Garmisch Semiconductor Conference); Vieweg Braunschweig: also *Rochester Semiconductor Conference*; to be published in *J. Phys. Chem. Solids*, 1959: *123*

— and BRIGGS, H. B. (1952a) *Phys. Rev.* **86** 647: *91, 143*

— — (1952b) *Phys. Rev.* **86** 697: *123*

— and WESTPHAL, W. C. (1956) *Phys. Rev.* **101** 1676: *91, 123*

HEAVENS, O. S. (1955) *Optical Properties of Thin Solid Films*; Butterworths, London: *14*

— and SMITH, S. D. (1957) *J. opt. Soc. Amer.* **47** 469: *14*

HENDERSON, S. T. (1958) *Brit. J. appl. Phys.* **9** 45: *98*

HENKELS, H. W. (1950) *J. appl. Phys.* **21** 725: *154*

— (1951) *J. appl. Phys.* **22** 916: *163*

— and MACZUK, J. (1953) *J. appl. Phys.* **24** 1056: *154, 164*

— — (1954) *J. appl. Phys.* **25** 1: *163*

HENVIS, B. W. *See* PICUS, G. S.

HERMAN, F. *See* BRAUNSTEIN, R.

— (1952) *Phys. Rev.* **88** 1210: *107*

— (1954) *Phys. Rev.* **93** 1214; **95** 847: *35, 113, 129*

— (1955a) *Proc. Inst. Radio Engrs, N.Y.* **43** 1703: *35, 50, 51, 113, 129, 150*

— (1955b) *J. Electronics* **1** 103: *50, 113, 129, 130, 227*

— (1957) *J. Phys. Chem. Solids* **2** 77: *107*

— GLICKSMAN, M. and PARMENTER, R. H. (1957) *Progress in Semiconductors* Vol. 2 1; Heywood, London: *20z*

HILL, J. E. and VAN VLIET, K. M. (1958) *J. appl. Phys.* **29** 177: *149*

HILSUM, C. *See* STARKIEWICZ, J.

— (1956) *Proc. phys. Soc., Lond.* **B69** 506: *152, 153, 154*

— (1957) *Proc. phys. Soc., Lond.* **B70** 1011: *225*

— and BARRIE, R. (1958) *Proc. phys. Soc., Lond.* **71** 676: *226, 227*

— OLIVER, D. J. and RICKAYSEN, G. (1955) *J. Electronics* **1** 134: *241*

— and ROSS, I. M. (1957) *Nature, Lond.* **179** 146: *241*

HOARD, J. L. *See* LAUBENGAYER, A. W.

— GELLER, S. and HUGHES, R. E. (1951) *J. Amer. chem. Soc.* **73** 1892: *102*

HOFMAN, D. *See* DE NOBEL, D.

— LELY, J. A. and VOLGER, J. (1957) *Physica* **23** 236: *112*

HÖHLER, G. (1949) *Ann. Phys., Lpz.* **4** 371: *216*

HOOGENSTRAATEN, W. *See* DIEMER, G.

HORN, F. H. *See* TAFT, E. A.

HOSLER, W. R. *See* BRECKENRIDGE, R. G.; FREDERIKSE, H. P. R.

HOUSTON, R. A. (1938) *Treatise on Light*; Longmans, London: *9*

HOWE, J. B. *See* CZYZAK, S. J.

HROSTOWSKI, H. J. *See* BRIGGS, H. B.

— and KAISER, R. H. (1957) *Phys. Rev.* **107** 966: *126*

— — (1958) *J. Phys. Chem. Solids* **4** 48, 315: *123, 125*

— MORIN, F. J., GEBALLE, T. H. and WHEATLEY, G. H. (1955) *Phys. Rev.* **100** 1672: *226, 230*

— WHEATLEY, G. H. and FLOOD, W. F. (1954) *Phys. Rev.* **95** 1683: *231, 232*

263

HUDSON, D. E. *See* SHAW, W. C.
HUGHES, R. E. *See* HOARD, J. L.
HUM, D. M. *See* BELL, D. G.
HUMPHREY, J. N. and PETRITZ, R. L. (1957*a*) *Phys. Rev.* **105** 1192: *198*
—— (1957*b*) *Phys. Rev.* **105** 1736: *198, 199*
— and SCANLON, W. W. (1957) *Phys. Rev.* **105** 469: *198, 199*
HUMPHREYS, D. L. O. *See* CHAMPION, F. C.
HUNTINGTON, H. B. and APKER, L. (1953) *Phys. Rev.* **89** 352: *98*
HURD, D. T. *See* LAUBENGAYER, A. W.
HUTSON, A. R. (1957) *Phys. Rev.* **108** 222: *202, 203*
— (1958) *Rochester Semiconductor Conference*; to be published in *J. Phys. Chem. Solids* 1959: *202*
HYMAN, R. A. (1956*a*) *Proc. phys. Soc., Lond.* **B69** 743: *163*
— (1956*b*) *Proc. phys. Soc., Lond.* **B69** 1085: *154, 165, 166*

IBUKI, S. and YOSHIMATSU, S. (1955) *J. phys. Soc. Japan* **10** 549: *48*
IRIE, T. *See* UNO, R.
IVEY, H. F. *See* DESTRIAU, G.

JACOBSMEYER, V. P. *See* BADER, L. J.
JAHODA, F. C. (1957) *Phys. Rev.* **107** 1261: *26*
JENKINS, D. P. (1954) *Physica* **20** 967: *113*
JENNY, D. A. and BUBE, R. H. (1954) *Phys. Rev.* **96** 1190: *202, 221*
— LOFERSKI, J. J. and RAPPAPORT, P. (1956) *Phys. Rev.* **101** 1208: *82, 122, 225*
JOHNSON, E. R. and CHRISTIAN, S. M. (1955) *Phys. Rev.* **95** 560: *150, 151*
JOHNSON, J. R. *See* MILLER, P. H.
JONES, D. A. *See* PAUL, W.
JONES, F. E. *See* SMITH, R. A.
JONES, H. *See* MOTT, N. F.
JONES, R. C. (1953) *J. opt. Soc. Amer.* **43** 1008: *198*
JONES, R. H. (1957) *Proc. phys. Soc., Lond.* **B70** 704 1025: *198*
JONES, R. V. *See* PAUL, W.

KAHN, A. H. (1955) *Phys. Rev.* **97** 1647: *120, 138, 140*
KAISER, R. H. *See* HROSTOWSKI, H. J.
KAISER, W. *See* FAN, H. Y.
— (1957) *Phys. Rev.* **105** 1751: *126*
— COLLINS, R. J. and FAN, H. Y. (1953) *Phys. Rev.* **91** 230 1380: *51, 139*
— and FAN, H. Y. (1954) *Phys. Rev.* **93** 911 977: *146, 148*
—— (1955) *Phys. Rev.* **98** 966: *232*
— KECK, P. H. and LANGE, C. F. (1956) *Phys. Rev.* **101** 1264: *126*
KANDARE, S. (1957) *C.R. Acad. Sci., Paris* **244** 571: *162*
KANE, E. O. (1957) *J. Phys. Chem. Solids* **1** 249: *227, 228, 234, 240*
KANEKO, H. *See* SATO, T.
KANEV, ST. *See* BORISSOV, M.
KAPLYANSKII, A. A. *See* GROSS, E. F.
KARRYEV, N. A. *See* GROSS, E. F.
KAZAN, B. and NICOLL, F. H. (1955) *Proc. Inst. Radio Engrs, N.Y.* **43** 1888: *98*
—— (1957) *J. opt. Soc. Amer.* **47** 887: *213*
KECK, P. H. *See* KAISER, W.
— (1952) *J. opt. Soc. Amer.* **42** 221: *152, 164, 165*
KEFFER, F. and PORTIS, A. M. (1957) *J. chem. Phys.* **27** 675: *202*
KENDALL, J. T. (1954) *Phil. Mag.* **45** 141: *179*

KEYES, R. J. *See* STEVENSON, D. T.; ZWERDLING, S.
— ZWERDLING, S., FONER, S., KOLM, H. H. and LAX, B. (1956) *Phys. Rev.* **104** 1804: *88, 233, 234*
KEYES, R. W. (1955) *Phys. Rev.* **99** 490: *230*
KIELY, D. G. *See* GEBBIE, H. A.
KIKOIN, I. K. and BYKOVSKII, YU. A. (1956) *Dokl. Akad. Nauk U.S.S.R.* **109** 735: *143*
— and NOSKOV, M. M. (1934) *Phys. Z. Soujet.* **5** 586: *63*
KIMBALL, G. E. (1935) *J. chem. Phys.* **3** 560: *43, 49, 108*
KIP, A. F. *See* DRESSELHAUS, G.; DEXTER, R. N.
— (1954) *Physica* **20** 813: *88*
KITTEL, C. *See* DRESSELHAUS, G.
— (1956) *Solid State Physics*, Wiley, N.Y.: *88*
KLICK, C. C. *See* LAMBE, J.
— (1953) *Phys. Rev.* **89** 274: *215, 216, 217, 220*
KLINGER, M. I. and CHABAN, M. M. (1956) *J. tech. Phys., Moscow* **26** 938: *87*
— and SAKER, E. W. (1953) *Proc. phys. Soc., Lond.* **B66** 1117: *158*
KLONTZ, E. E. *See* FAN, H. Y.
KOC, S. (1957) *Czech. J. Phys.* **7** 91: *136, 141, 238*
KOESTER, C. J. *See* GIVENS, M. P.
KOHN, W. (1955) *Phys. Rev.* **98** 1856: *126*
— (1957) *Solid State Physics* **5** 257; Academic Press, N.Y.: *123, 124, 125, 126, 130, 145*
KOHNKE, E. E. *See* EWALD, A. W.
— and EWALD, A. W. (1956) *Phys. Rev.* **102** 1481: *179*
KOLM, C., KULIN, S. A. and AVERBACH, B. L. (1957) *Phys. Rev.* **108** 965: *225*
KOLM, H. H. *See* KEYES, R. J.; ZWERDLING, S.
KOLOMOETS, N. V., STAVITSKAYA, T. S. and STIL'BANS, L. S. (1957) *J. tech. Phys., Moscow* **27** 73: *182*
KOMAR, A. P., REINOV, N. M. and SHALYT, S. S. (1954) *Dokl. Akad. Nauk U.S.S.R.,* **96** 47: *63*
KOROLEV, N. V. *See* GALKIN, L. N.
KOZYREV, P. T. (1957) *J. tech. Phys., Moscow* **27** 35: *163*
KRISHNAN, R. S. (1946) *Proc. Indian Acad. Sci.* **A24** 33: *110*
KROEMER, H. (1957) *Brooklyn Polytechnic Symposia* **7** 143; Interscience, N.Y.: *150*
KRÖGER, F. A. *See* BLOEM, J.; LELY, J. A.
— and DE NOBEL, D. (1955) *J. Electronics* **1**, 190: *221*
— VINK, H. J. and VOLGER, J. (1954) *Physica* **20** 1095: *217, 218*
KRONIG, R. DE L. (1931) *Proc. Roy. Soc.* **133** 245: *29*
KULIN, S. A. *See* KOLM, C.
KURNICK, S. W. *See* MITCHELL, G. R.
— STRAUSS, A. J. and ZITTER, R. N. (1954) *Phys. Rev.* **94** 1791: *63*
— and ZITTER, R. N. (1956a) *J. appl. Phys.* **27** 278: *63, 69, 72*
— — (1956b) *Photoconductivity Conference*, Wiley, N.Y.: *63, 73, 230, 241*
KUSS, E. (1956) *Chem.-Ing.-Tech.* **28** 141: *107*
KYRYASHKIN, V. I. *See* KYRYASHKINA, E. I.
KYRYASHKINA, E. I., POPOV, F. M., BILENKO, D. I. and KYRYASHKIN, V. I. (1957) *J. tech. Phys., Moscow* **27** 85: *204*

LAGRENAUDIE, J. (1953a) *J. chim. Phys.* **50** 629: *102, 104, 105*
— (1953b) *J. Phys. Radium* **14** 14: *102, 103*
— (1956) *Ann. Télécomm.* **11** 127 132: *63*

LAMBE, J. (1955) *Phys. Rev.* **98** 985: *219*

— and KLICK, C. C. (1958) *Progress in Semiconductors* Vol. 3 185; Heywood, London: *214*

LANDSBERG, P. T. (1957) *Proc. phys. Soc., Lond.* **B70** 1175: *238*

— (1958) *Proc. phys. Soc., Lond.* **71** 69: *229*

— and Moss, T. S. (1956) *Proc. phys. Soc., Lond.* **B69** 661: *92, 238*

LANGE, C. F. *See* KAISER, W.

LARACH, S., SHRADER, R. E. and STOCKER, C. F. (1957) *Phys. Rev.* **108** 587: *203*

LARK-HOROWITZ, K. *See* FAN, H. Y.

LASHKAREV, V. E., MEDVEDEV, V. C., SKOPENKO, A. I., FEDORUS, G. A. and SHEINKMAN, M. K. (1952) *Dokl. Akad. Nauk U.S.S.R.* **86** 905: *218*

LASSER, M. E., CHOLET, P. and WURST, E. C. (1958) *J. opt. Soc. Amer.* **48** 468: *148, 240*

— and LEVINSTEIN, H. (1954) *Phys. Rev.* **96** 47: *185, 190*

LAUBENGAYER, A. W., HURD, D. T., NEWKIRK, A. E. and HOARD, J. L. (1943) *J. Amer. chem. Soc.* **65** 1924: *102*

LAUTZ, G. *See* APPEL, J.

LAWRANCE, R. (1951) *Aust. J. sci. Res.* **A4** 569: *194*

LAWSON, W. D. *See* AVERY, D. G.

— (1951) *J. appl. Phys.* **22** 1444: *181*

— (1952) *J. appl. Phys.* **23** 495: *181*

— NEILSON, S. and YOUNG, A. S. (1958) *Brussels Solid. State Congress*; Academic Press, N.Y.: *204, 222*

LAX, B. *See* DEXTER, R. N.; KEYES, R. J.; ROTH, L. M.; ZEIGER, H. J.; ZWERDLING, S.

— (1958) *Rev. mod. Phys.* **30** 122: *46, 88*

— and ROTH, L. M. (1955) *Phys. Rev.* **98** 5 549: *87*

— ZEIGER, H. J. and DEXTER, R. N. (1954) *Physica* **20** 818: *88*

LAX, M. and BURSTEIN, E. (1955a) *Phys. Rev.* **97** 39: *141*

— — (1955b) *Phys. Rev.* **100** 592: *125*

LEHOVEC, K. (1952) *Proc. Inst. Radio Engrs, N.Y.* **40** 1407: *29, 140*

LEIVO, W. J. and SMOLUCHOWSKI, R. (1955) *Phys. Rev.* **98** 1532: *111*

LELY, J. A. *See* HOFMAN, D.

— and KRÖGER, F. A. (1956/8) *Halbleiter und Phosphore* (Garmisch Semiconductor Conference); Vieweg, Braunschweig: *112*

LEMPICI, A., FRANKL, D. R. and BROPHY, V. A. (1957) *Phys. Rev.* **107** 1238: *213*

LEVINSTEIN, H. *See* LASSER, M. E.

— (1956) *Photoconductivity Conference*; Wiley, N.Y.: *185, 190, 191, 196, 198*

LEVITAS, A., WANG, C. C. and ALEXANDER, B. H. (1954) *Phys. Rev.* **95** 846: *150*

LEVY, J. L. (1953) *Phys. Rev.* **92** 215: *199*

LIDIARD, A. B. *See* STEPHEN, M. J.

LILBURN, M. T. *See* REYNOLDS, W. N.

LIND, E. L. *See* BUBE, R. H.

LIPSON, H. G. *See* BURSTEIN, E.

LIZELL, B. (1952) *J. chem. Phys.* **20** 672: *154*

LOFERSKI, J. J. *See* HARTIG, P. A.; JENNY, D. A.

— (1954) *Phys. Rev.* **93** 707: *164, 165, 167, 169, 170, 175, 176, 177*

— (1956) *J. appl. Phys.* **27** 777: *82*

LONG, D. (1955) *Phys. Rev.* **99** 388: *230*

— (1956) *Phys. Rev.* **101** 1256: *167*

— and Miller, P. H. (1955) *Phys. Rev.* **98** 1192: *230*

LUDWIG, G. W. and WATTERS, R. L. (1956) *Phys. Rev.* **101** 1699: *114*

LUKES, F. (1956) *Czech. J. Phys.* **6** 359: *225*

266

LUMMIS, F. L. and PETRITZ, R. L. (1957) *Phys. Rev.* **105** 502: *198, 200*
LUTTINGER, J. M. (1956) *Phys. Rev.* **102** 1030: *89*

MACFARLANE, G. G., MCLEAN, T. P., QUARRINGTON, J. E. and ROBERTS, V. (1957)
    Phys. Rev. **108** 1377: *48, 130, 135, 137*
—— —— —— (1958) *Proc. phys. Soc., Lond.* **71** 863: *48*
— and ROBERTS, V. (1955a) *Phys. Rev.* **97** 1914: *134, 137*
—— —— (1955b) *Phys. Rev.* **98** 1865: *37, 117, 118*
—— —— (1956/8) *Halbleiter und Phosphore* (Garmisch Semiconductor Conference);
    Vieweg, Braunschweig: *143*
MACKINTOSH, I. M. (1956) *Proc. phys. Soc., Lond.* **B69** 115: *193, 194, 195*
— and ALLEN, J. W. (1955) *Proc. phys. Soc., Lond.* **B68** 985: *93*
MACZUK, J. *See* HENKELS, H. W.
MAHLMAN, G. W. (1956) *Phys. Rev.* **103** 1619: *200*
MAITA, J. P. *See* MORIN, F. J.
MALKOVSKA, M. *See* DRAHOKOUPIL, J.
MARGENAU, H. *See* SAMPSON, D.
MARTIN, A. E. *See* ROBERTSON, R.
MASLAKOVETS, YU. P., POLTINNIKOV, S. A., DUBROVSKI, G. B. and SUBASHIEV, V. K.
    (1956) *J. tech. Phys., Moscow* **26** 2396: *122*
MATTHEWS, I. G. *See* DYER, H. B.
MCCLYMONT, D. R. *See* AUSTIN, I. G.
MCKAY, K. G. *See* CHYNOWETH, A. G.
MCKIM, F. S. *See* BUCK, T. M.
MCLEAN, T. P. *See* MACFARLANE, G. G.
MCSKIMIN, H. J. and BOND, W. L. (1957) *Phys. Rev.* **105** 116: *108*
MEDVEDEV, V. C. *See* LASHKAREV, V. E.
MEIER, W. (1910) *Ann. Phys., Lpz.* **31** 1017: *158*
MERZ, W. (1958) *Brussels Solid State Congress*; Academic Press, N.Y.: *213*
MILLER, P. H. *See* LONG, D.
— (1951) *Semiconducting Materials* (Reading Conference); Butterworths, London:
    203, 205
— and JOHNSON, J. R. (1954) *Physica* **20** 1026: *9*
MINDEN, H. T. *See* HARADA, R. H.
MITCHELL, E. J. W. (1955) *Proc. phys. Soc., Lond.* **B68** 973: *86*
MITCHELL, G. R. *See* GOLDBERG, A. E.
— and GOLDBERG, A. E. (1953) *Phys. Rev.* **93** 1421: *194*
—— and KURNICK, S. W. (1955) *Phys. Rev.* **97** 239: *240*
MIYASAWA, H. and SUGAIKE, S. (1954) *J. phys. Soc. Japan* **9** 648: *221*
MOCHAN, I. V. (1956) *J. tech. Phys., Moscow* **25** 1003: *174*
MOGLICH, F. and ROMPE, R. (1942) *Z. Phys.* **119** 472: *209*
MOLLWO, E. (1956) *Photoconductivity Conference*, Wiley, N.Y.: *202, 203*
— and STOCKMAN, F. (1948) *Ann. Phys., Lpz.* **3** 223: *206*
MOORE, A. R. *See* BRAUNSTEIN, R.
MOOSER, E. *See* BUSCH, G.
— and PEARSON, W. B. (1956) *Canad. J. Phys.* **12A** 1369: *175*
—— —— (1957) *J. Electronics* **2** 406: *179*
MORIN, F. J. *See* HROSTOWSKI, H. J.
— (1954) *Phys. Rev.* **93** 62: *131*
— and MAITA, J. P. (1954a) *Phys. Rev.* **94** 1525: *131*
—— —— (1954b) *Phys. Rev.* **96** 28: *115*
MORITA, N. (1954) *J. sci. Res. Instrum.* (*Tokyo*) **48** 8: *103, 104, 105*

MOSER, F. and URBACH, F. (1956) *Phys. Rev.* **102** 1519: *39*

MOSS, T. S. *See* AVERY, D. G.; LANDSBERG, P. T.; SMITH, S. D.; WALTON, A. K,

— (1949a) *Proc. phys. Soc., Lond.* **A62** 264: *175*

— (1949b) *Proc. phys. Soc., Lond.* **B62** 743: *196*

— (1950) *Proc. phys. Soc., Lond.* **B63** 167: *225*

— (1951) *Proc. phys. Soc., Lond.* **A64** 1055: *197*

— (1952a) *Photoconductivity in the Elements*; Butterworths, London: *17, 39, 40, 45, 48, 49, 53, 98, 102, 103, 104, 105, 107, 108, 111, 118, 141, 142, 153, 156, 159, 161, 167, 169, 170, 175, 176, 177, 179, 180, 181, 251*

— (1952b) *Proc. phys. Soc., Lond.* **B65** 62: *167*

— (1953a) *Proc. phys. Soc., Lond.* **B66** 993: *63, 66, 69, 72, 192, 193*

— (1953b) *Proc. phys. Soc., Lond.* **B66** 141: *24*

— (1953c) *Proc. phys. Soc., Lond.* **B66** 750: *142*

— (1953d) *Research, Lond.* **6** 259: *197*

— (1954a) *Proc. phys. Soc., Lond.* **B67** 775: *30, 40, 45, 46, 231*

— (1954b) *Photoconductivity Conference*: Wiley, N.Y.: *30, 230, 241*

— (1955a) *J. Electronics* **1** 126: *58*

— (1955b) *Proc. Inst. Rad. Engrs, N.Y.* **43** 1869: *186, 195*

— (1955c) *Proc. phys. Soc., Lond.* **B68** 697: *181*

— (1956/8) *Halbleiter und Phosphore* (Garmisch Semiconductor Conference); Vieweg, Braunschweig: *63*

— (1957a) *Proc. phys. Soc., Lond.* **B70** 247: *93, 95*

— (1957b) *Proc. phys. Soc., Lond.* **B70** 778/9: *42, 237*

— (1958a) *Advances in Spectroscopy* Vol. 1; Interscience, N.Y.: *195, 241*

— (1958b) *Brooklyn Polytechnic Symposium of 1957*, Vol. 7; Interscience, N.Y.: *195, 226*

— (1958c) *Rochester Semiconductor Conference*; to be published in *J. Phys. Chem. Solids*, 1959: *46*

— and HAWKINS, T. D. H. (1956) *J. Phys. Radium* **17** 712: *91*

— — (1958a) *Proc. phys. Soc., Lond.* **72** 270: *101, 238, 239*

— — (1958b) *Phys. Rev. Letters* **1** 129: *18*

— — and SMITH, S. D. (1956) *Semiconductor Conference (Rugby) Rep.*; Phys. Soc., (London): *91, 97*

— PINCHERLE, L. and WOODWARD, A. M. (1953) *Proc. phys. Soc., Lond.* **B66** 743: *63, 67, 68, 74*

— and SMITH, S. D. (1958) *Brussels Solid State Congress*; Academic Press, N.Y.: also *Rochester Semiconductor Conference*, to be published in *J. Phys. Chem. Solids*, 1959: *234, 240*

— — and HAWKINS, T. D. H. (1957); *Proc. phys. Soc., Lond.* **B70** 776: *13, 41, 96, 228, 235*

MOTT, N. F. and GURNEY, R. W. (1948) *Electronic Processes in Ionic Crystals*, 2nd Edn; Clarendon Press, Oxford: *17, 28*

— and JONES, H. (1936) *Properties of Metals and Alloys*, Oxford University Press: *16, 43*

MUNSCH, M. (1954) *Rev. Opt. (théor. instrum.)* **10** 497: *195*

NARASIMHAN, P. T. (1955) *Proc. phys. Soc., Lond.* **B68** 315: *109*

NEAL, F. A. *See* CUSTERS, J. F. H.

NEILSON, S. *See* LAWSON, W. D.

NEUMARK, G. F. (1956) *Phys. Rev.* **103** 41: *214*

NEURINGER, L. J. (1955) *Phys. Rev.* **98** 1185 1193: *169*

NEWKIRK, A. E. *See* LAUBENGAYER, A. W.

NEWMAN, R. *See* DASH, W. C.
— (1953) *Phys. Rev.* **91** 1313: *91, 143, 144*
— (1954) *Phys. Rev.* **94** 278 1530: *147, 148* —
— (1955) *Phys. Rev.* **99** 465: *127*
— (1956) *Phys. Rev.* **103** 103: *125*
— (1957) *Phys. Rev.* **105** 1715: *145*
— and TYLER, W. W. (1954) *Phys. Rev.* **94** 1419; **96** 882: *148, 149*
— — (1957) *Phys. Rev.* **105** 885: *140*
— WOODBURY, H. H. and TYLER, W. W. (1956) *Phys. Rev.* **102** 613: *148, 149*
NICOLL, F. H. *See* KAZAN, B.
NOBEL, D. DE. *See* DE NOBEL.
NOSKOV, M. M. *See* KIKOIN, I. K.
NOVIKOV, B. V. *See* GROSS, E. F.
NOZIÈRES, P. and PINES, D. (1958) *Phys. Rev.* **109** 741 762 1062: *16, 17*
NUSSBAUM, A. (1953) *Pennsylvania University Report* (largely published as NUSSBAUM, A., 1954): *176*
— (1954) *Phys. Rev.* **94** 337: *164, 165, 167, 169*

OBERLY, J. J. (1954) *Phys. Rev.* **93** 911: *63*
OETJEN, R. A. *See* YOSHINAGA, H.
OLIVER, D. J. *See* HILSUM, C.
— (1957) *Proc. phys. Soc., Lond.* **B70** 331: *241*
ORTHUBER, R. K. and ULLERY, L. R. (1954) *J. opt. Soc. Amer.* **44** 297: *221*
OSHINSKY, W. *See* BRECKENRIDGE, R. G.
OSWALD, F. (1955) *Z. Naturf.* **10a** 927: *230*
OTSUKA, E. *See* TOMURA, M.

PARMENTER, R. H. *See* HERMAN, F.
— (1955) *Phys. Rev.* **97** 587: *40, 150*
PATRICK, L. *See* CHOYKE, W. J.
PAUL, W. *See* CARDONA, M.; WARSCHAUER, D. M.
— JONES, D. A. and JONES, R. V. (1951) *Proc. phys. Soc., Lond.* **B64** 528: *183, 185*
— and JONES, R. V. (1953) *Proc. phys. Soc., Lond.* **B66** 194: *183*
— and PEARSON, G. L. (1955) *Phys. Rev.* **98** 1755: *118, 119*
PEARLSTEIN, E. A. and SUTTON, R. B. (1950) *Phys. Rev.* **79** 907: *108*
PEARSON, G. L. *See* CHAPIN, D. M.; PAUL, W.
— and TANENBAUM, M. (1953) *Phys. Rev.* **90** 153: *46, 227*
PEARSON, W. B. *See* MOOSER, E.
PEIERLS, R. E. (1955) *Quantum Theory of Solids*; Clarendon Press, Oxford: *47*
PENSAK, L. (1958) *Phys. Rev.* **109** 601: *222*
PEPPER, R. R. *See* FAN, H. Y.
PETRITZ, R. L. *See* HUMPHREY, J. N.; LUMMIS, F. L.; SCANLON, W. W.
— (1956) *Phys. Rev.* **104** 1508: *200*
— and SCANLON, W. W. (1955) *Phys. Rev.* **97** 1620: *186, 187*
PFANN, W. G. (1957) *Solid State Physics* **4** 424; Academic Press, N.Y.: *113, 129*
— and VAN ROOSBROECK, W. (1954) *J. appl. Phys.* **25** 1422: *82*
PFEIFFER, H. U. (1956) *Z. Naturf.* **11a** 164: *195*
PFESTORF, G. (1926) *Ann. Phys., Lpz.* **81** 906: *118*
PFISTER, H. (1956) *Z. Naturf.* **11a** 434: *77*
PHILLIPS, J. *See* BLOUNT, E.
PICUS, G. S. *See* BURSTEIN, E.
— BURSTEIN, E. and HENVIS, B. W. (1957) *Bull. Amer. phys. Soc.* II **2** 66: *225*

REFERENCE LIST AND AUTHOR INDEX

PINCHERLE, L. *See* BELL, D. G.; MOSS, T. S.
— (1955) *Proc. phys. Soc., Lond.* **B68** 319: *193*
— (1956) *Photoconductivity Conference*; Wiley, N.Y.: *63, 75*
— and RADCLIFFE, J. M. (1956) *Phil. Mag. Suppl.* **5** 271: *223*
PINES, D. *See* NOZIÈRES, P.
PIPER, W. W. (1953) *Phys. Rev.* **92** 23: *39, 202, 206, 207, 208, 212, 213*
— and WILLIAMS, F. E. (1958) *Solid State Physics* **6** 96; Academic Press, N.Y.: *98, 214*
PIZZARELLO, F. A. *See* DE VAUX, L. H.
PLESSNER, K. W. (1951) *Proc. phys. Soc., Lond.* **B64** 681: *163*
POLTINNIKOV, S. A. *See* MASLAKOVETS, YU. P.
POPOV, F. M. *See* KYRYASHKINA, E. I.
PORTIS, A. M. *See* KEFFER, F.
POTTER, R. F. (1956a) *Phys. Rev.* **103** 861: *228*
— (1956b) *Phys. Rev.* **103** 47: *230, 231*
— (1957a) *Phys. Rev.* **108** 652: *227*
— (1957b) *J. Phys. Chem. Solids* **3** 223: *235*
POWELL, J. A. *See* CRESSWELL, I. G.
PRATT, R. G. *See* SMOLLETT, M.
PRIBYTKOVA, N. N. (1957) *Optika i Spektrosk* **2** 623: *162*
PRICE, W. C. *See* ROBINSON, T. S.
PRINCE, M. B. (1955) *J. appl. Phys.* **26** 534: *82, 122*
— and WOLF, M. (1958) *Brussels Solid State Congress*; Academic Press, N.Y.: *82, 122*
PRIVKOWSKI, T. *See* FELTYNOWSKI, A.
PUTLEY, E. H. (1955) *Proc. phys. Soc., Lond.* **B68** 22 35: *186*

QUARRINGTON, J. E. *See* MACFARLANE, G. G.; ROBERTS, V.

RAAL, F. A. (1958) *Proc. phys. Soc., Lond.* **71** 846: *112*
RABENHORST, H. *See* ECKART, F.
RACETTE, J. H. (1957) *Phys. Rev.* **107** 1542: *112*
RADCLIFFE, J. M. *See* PINCHERLE, L.
RADHAKRISHNAN, T. *See* RAMACHANDRAN, G. N.
RAMACHANDRAN, G. N. (1947a) *Proc. Indian Acad. Sci.* **A25** 206: *109*
— (1947b) *Proc. Indian Acad. Sci.* **A25** 375: *210*
— and RADHAKRISHNAN, T. (1952) *Phil. Mag.* **43** 317: *109*
RAMAN, C. V. (1956) *Proc. Indian Acad. Sci.* **44** 99: *107*
RAMDAS, A. K. *See* FAN, H. Y.
RANK, D. H., BENNETT, H. E. and CRONEMEYER, D. C. (1954) *J. opt. Soc. Amer.* **44** 13: *131, 132, 133*
RAPPAPORT, P. *See* JENNY, D. A.
— (1954) *Phys. Rev.* **93** 246: *82*
RASHBA, E. I. *See* BROUDE, V. L.
RAU, R. R. and CASPARI, M. E. (1955) *Phys. Rev.* **100** 632: *87, 131*
REDFIELD, A. G. (1954) *Phys. Rev.* **94** 526: *107*
REDFIELD, D. (1955a) *Phys. Rev.* **100** 1094: *176, 177, 178*
— (1955b) *Technical Report No. 13*; University of Pennsylvania, Pa.: *177*
— (1956) *Photoconductivity Conference*; Wiley, N.Y.: *177*
REINOV, N. M. *See* KOMAR, A. P.
REITZ, J. R. (1957) *Phys. Rev.* **105** 1233: *174, 175*
RENNIE, A. E. *See* AVERY, D. G.

270

REYNOLDS, C. C. *See* CZYZAK, S. J.; REYNOLDS, D. C.
REYNOLDS, D. C. *See* CZYZAK, S. J.
— and CZYZAK, S. J. (1954) *Phys. Rev.* **96** 1705: *221*
— — ALLEN, R. C. and REYNOLDS, C. C. (1955) *J. opt. Soc. Amer.* **45** 136: *214,*
    217, 219
REYNOLDS, W. N., LILBURN, M. T. and DELL, R. M. (1958) *Proc. phys. Soc., Lond.*
    71 416: *225*
RICHTER, H., BERCKHEMER, H. and BREITLING, G. (1954) *Z. Naturf.* **9a** 236: *180*
RICKAYSEN, G. *See* HILSUM, C.
RIDEAL, E. K. *See* TAYLOR, A. M.
▶RITTNER, E. S. (1956) *Photoconductivity Conference*; Wiley, N.Y.: *55, 82, 200*
— and FINE, S. (1955) *Phys. Rev.* **98** 545: *200*
ROBERTS, S. (1955) *Phys. Rev.* **100** 1667: *30*
ROBERTS, V. *See* MACFARLANE, G. G.
— and QUARRINGTON, J. E. (1955) *J. Electronics* **1** 152: *37, 38, 228, 230*
ROBERTSON, R., FOX, J. J. and MARTIN, A. E. (1934) *Phil. Trans.* **A232** 463: *109*
ROBINSON, T. S. and PRICE, W. C. (1953) *Proc. phys. Soc., Lond.* **B66** 969: *26*
ROMPE, R. *See* MOGLICH, F.
ROOSBROECK, W. VAN. *See* VAN ROOSBROECK, W.
ROSE, A. *See* RUPPEL, W.
— (1955) *Proc. Inst. Radio Engrs, N.Y.* **43** 1850: *53, 205, 220*
— (1957) *Progress in Semiconductors* Vol. 2; Heywood, London: *220*
ROSENFELD, L. (1951) *Theory of Electrons*; North Holland Publishing Co.,
    Amsterdam: *16*
ROSS, I. M. *See* HILSUM, C.
ROSSEL, J. *See* ALLEMAND, C.
ROTH, L. M. *See* LAX, B.; ZWERDLING, S.
— LAX, B. and ZWERDLING, S. (1958) to be published in *Phys. Rev.*: *89*
ROUX, J. (1956) *Ann. Phys. Paris* **1** 493: *206*
RUPPEL, W., GERRITSEN, H. J. and ROSE, A. (1957) *Helv. phys. acta* **30** 495: *202*

SAKER, E. W. *See* CUNNELL, F. A.; GEBBIE, H. A.; KLINGER, M. I.
— (1952) *Proc. phys. Soc., Lond.* **B65** 785: *153, 154, 156, 157*
SALZBERG, C. D. and VILLA, J. J. (1957) *J. opt. Soc. Amer.* **47** 244: *115, 131, 132*
— — (1958) *J. opt. Soc. Amer.* **48** 579: *131*
SAMPSON, D. and MARGENAU, H. (1956) *Phys. Rev.* **103** 879: *125*
SATO, T. and KANEKO, H. (1949) *Tech. Rep. Tôhoku Univ.* **14** 45: *155*
SCANLON, W. W. *See* BREBRICK, R. F.; HUMPHREY, J. N.; PETRITZ, R. L.
— (1953) *Phys. Rev.* **92** 1573: *181*
— (1957) *Phys. Rev.* **106** 718: *193*
— (1958*a*) *Phys. Rev.* **109** 47: *183, 185, 190, 193*
— (1958*b*) *Rochester Semiconductor Conference*; to be published in *J. Phys. Chem.*
    Solids 1959: *185, 193*
— and BREBRICK, R. F. (1954) *Physica* **20** 1090: *181*
— — and PETRITZ, R. L. (1956) *Photoconductivity Conference*; Wiley, N.Y.: *181*
SCHMIDT, L. A. (1953) *Phys. Rev.* **92** 1373: *111*
SCHMITZ, W. D. *See* SUITS, G. H.
SCHNEIDER, M. *See* BUSCH, G.
SCHOTTKY, W. (1953) *Z. Naturf.* **8a** 457: *156*
SCHWARZ, E. (1951) *Proc. phys. Soc., Lond.* **B64** 821: *207*
SCHWETZOFF, V. (1957) *C.R. Acad. Sci., Paris* **245** 149: *198*
SCIAMA, D. W. *See* BELL, D. G.

SCLAR, N. *See* BURSTEIN, E.
— and BURSTEIN, E. (1955) *Phys. Rev.* **98** 1757: *123, 149*
SEITZ, F. *See* BARNES, R. B.
— (1940) *Modern Theory of Solids*; McGraw Hill, N.Y.: *16*
SHAKIN, C. and BIRMAN, J. (1958) *Phys. Rev.* **109** 818: *207*
SHALYT, S. S. *See* KOMAR, A. P.
— (1956) *Dokl. Akad. Nauk U.S.S.R.* **109** 750: *174*
SHAW, W. C., HUDSON, D. E. and DANIELSON, G. C. (1953) *Phys. Rev.* **91** 208: *102*
SHEINKMAN, M. K. *See* LASHKAREV, V. E.
SHEPHERD, M. L. *See* FAN, H. Y.
SHINOHARA, K. *See* UNO, R.
SHNEIDER, A. D. (1956) *J. tech. Phys., Moscow* **26** 1436: *206*
SHOCKLEY, W. *See* BARDEEN, J.; VAN ROOSBROECK, W.
— (1950) *Electrons and Holes in Semiconductors*; Van Nostrand, N.Y.: *44, 78, 155*
SHOGENJI, K. and UCHIYAMA, S. (1957) *J. phys. Soc., Japan* **12** 252 431: *181*
SHRADER, R. E. *See* LARACH, S.
SHUBA, Y. A. (1956) *J. tech. Phys., Moscow* **26** 1129: *221*
SHULMAN, R. G. (1957) *J. Phys. Chem. Solids* **2** 115: *124*
SIEGMUND, W. P. *See* GIVENS, M. P.
SIMON, I. (1951) *J. opt. Soc. Amer.* **41** 336: *16*
SINTON, W. M. and DAVIS, W. C. (1954) *J. opt. Soc. Amer.* **44** 503: *211*
SKINNER, H. W. B. (1938) *Rep. Progr. Phys.* **5** 271: *43*
SKOPENKO, A. I. *See* LASHKAREV, V. E.
SLATER, J. C. (1956) *Phys. Rev.* **103** 1631: *200*
SMITH, R. A. (1953) *Phil. Mag. Suppl.* **2** 321: *181*
— (1954) *Physica* **20** 910: *181, 187, 188*
— JONES, F. E. and CHASMAR, R. P. (1957) *Detection and Measurement of Infra-Red Radiation*; Clarendon Press, Oxford: *185*
SMITH, S. D. *See* HEAVENS, O. S.; MOSS, T. S.
— and MOSS, T. S. (1958) *J. sci. Instrum.* **35** 105: *226*
SMOLLETT, M. and PRATT, R. G. (1955) *Proc. phys. Soc., Lond.* **B68** 390: *199*
SMOLUCHOWSKI, R. *See* LEIVO, W. J.
SOBOLEV, V. V. *See* GROSS, E. F.
SOCHARD, I. *See* SOMMERS, H. S.
SOMMERS, H. S., BERRY, R. E. and SOCHARD, I. (1956) *Phys. Rev.* **101** 987: *69, 220*
SOSNOWSKI, L. *See* CHECINSKA, H.
— (1957) *Phys. Rev.* **107** 1193: *141*
SPEAR, W. E. (1957) *Proc. phys. Soc., Lond.* **B70** 669: *155, 159, 160, 163*
SPENSER, H. E. (1958) *Phys. Rev.* **109** 1074: *199*
SPITZER, W. G. *See* FAN, H. Y.
— and FAN, H. Y. (1955) *Phys. Rev.* **99** 1893: *235*
— — (1957*a*) *Phys. Rev.* **106** 268: *120, 121*
— — (1957*b*) *Phys. Rev.* **106** 882: *14, 30, 33, 114, 132, 138, 234, 237, 238*
STARKIEWICZ, J., BATE, G., BENNETT, H. and HILSUM, C. (1957) *Proc. phys. Soc., Lond.* **B70** 258: *194*
STAVITSKAYA, T. S. *See* KOLOMOETS, N. V.
STEGMAN, H. (1957) *Naturwissenschaften* **44** 108: *160*
STEIBER, A. *See* FREYMANN, R.
STEPHEN, M. J. (1958*a*) *Proc. Camb. Phil. Soc.* **54** 81: *87*
— (1958*b*) *Proc. phys. Soc., Lond.* **71** 485: *112*
— and LIDIARD, A. B. (1958/9) *J. Phys. Chem. Solids*; in press: *86*
STERN, F. *See* TALLEY, R. M.

STERN, F. and TALLEY, R. M. (1957) *Phys. Rev.* **108** 158: *228*
STEVENSON, D. T. and KEYES, R. J. (1954) *Phys. Rev.* **94** 1416: *143*
STIL'BANS, L. S. *See* KOLOMOETS, N. V.
STOCKER, C. F. *See* LARACH, S.
STOCKMAN, F. *See* MOLLWO, E.
STRAKHOV, L. P. *See* BERLAGA, R. YA.
STRATTON, J. A. (1941) *Electromagnetic Theory*; McGraw Hill, N.Y.: *4*
STRATTON, R. *See* CHASMAR, R. P.
STRAUMARIS, M. E. (1940) *Z. Kristallogr.* **102** 432: *152*
— and AKA, E. Z. (1952) *J. appl. Phys.* **23** 330: *129*
STRAUSS, A. J. *See* KURNICK, S. W.
STUKE, J. (1953) *Z. Phys., Lpz.* **134** 194: *152, 153, 154, 160, 161, 162*
— (1954) *Z. Phys., Lpz.* **137** 401: *203*
SUBASHIEV, V. K. *See* MASLAKOVETS, YU. P.
SUGAIKE, S. *See* MIYASAWA, H.
SUGITA, T. *See* YAKAMA, E.
SUITS, G. H., SCHMITZ, W. D. and TERHUNE, R. W. (1956) *J. appl. Phys.* **27** 1385: *241*
SUTTON, R. B. *See* PEARLSTEIN, E. A.

TAFT, E. *See* APKER, L.
— and APKER, L. (1952) *J. chem. Phys.* **20** 1648: *98*
— — (1954) *Phys. Rev.* **96** 1946: *179*
TAFT, E. A. and HORN, F. H. (1954) *Phys. Rev.* **93** 64: *124*
TALLEY, R. M. *See* STERN, F.
— and ENRIGHT, D. P. (1954) *Phys. Rev.* **95** 1092: *225*
— and STERN, F. (1955) *J. Electronics* **1** 186: *46*
TANENBAUM, M. *See* BRIGGS, H. B.; PEARSON, G. L.
— and BRIGGS, H. B. (1953) *Phys. Rev.* **91** 1561: *230, 231*
TANUMA, S. (1954) *Sci. Rep. Res. Insts. Tôhoku Univ.* **A6** 159: *167, 170, 172*
TAUC, J. *See* DRAHOKOUPIL, J.
— (1954) *Czech. J. Phys.* **4** 311: *142*
— (1955) *Czech. J. Phys.* **5** 178 301: *58*
— (1956) *Czech. J. Phys.* **6** 421: *74*
— (1958) *Rochester Semiconductor Conference*; to be published in *J. Phys. Chem. Solids,* 1959: *238*
— and ABRAHAM, A. (1954) *Czech. J. Phys.* **4** 478: *230*
TAYLOR, A. M. and RIDEAL, E. K. (1927) *Proc. Roy. Soc.* **A115** 589: *156*
TERHUNE, R. W. *See* SUITS, G. H.
TOMAN, L. *See* WOLFE, G. A.
TOMBOULIAN, D. H. and BEDO, D. E. (1956) *Phys. Rev.* **104** 590: *43, 130*
TOMURA, M. and OTSUKA, E. (1955) *J. phys. Soc., Japan* **10** 313: *137*
TORUN, T. *See* FELTYNOWSKI, A.
TREU, M. *See* FISCHER, F.
— (1939) Unpublished report of German wartime work: *206*
TUFTE, O. N. *See* EWALD, A. W.
TURNER, W. J. *See* BURSTEIN, E.
TUZZOLINO, A. J. (1957) *Phys. Rev.* **105** 1411: *227*
— (1958) *Phys. Rev.* **109** 1980: *227*
TYLER, W. W. *See* NEWMAN, R.; WOODBURY, H. H.
— (1958) *Rochester Semiconductor Conference*; to be published in *J. Phys. Chem. Solids* 1959: *145*

UCHIYAMA, S. *See* SHOGENJI, K.
ULLERY, L. R. *See* ORTHUBER, R. K.
UNO, R., IRIE, T., YOSHIDA, S. and SHINOHARA, K. (1953) *J. sci. Res. Inst. Tokyo* **47** 216: *102*
URBACH, F. *See* MOSER, F.
— (1953) *Phys. Rev.* **92** 1324: *39*

VAN DOORN, C. Z. (1954) *Physica* **20** 1155: *209*
— and DE NOBEL, D. (1956) *Physica* **22** 338: *221*
VAN DYKE, G. D. (1922) *J. opt. Soc. Amer.* **6** 917: *170*
VAN ROOSBROECK, W. *See* PFANN, W. G.
— (1956) *Phys. Rev.* **101** 1713: *67, 68, 74*
— and SHOCKLEY, W. (1954) *Phys. Rev.* **94** 1558: *92, 98*
VAN VLIET, K. M. *See* HILL, J. E.
VAUX, L. H. DE. *See* DE VAUX, L. H.
VERNIER, P. (1953) *J. Phys. Radium* **14** 175: *185*
VILLA, J. J. *See* SALZBERG, C. D.
VINK, H. J. *See* KRÖGER, F. A.
VOLGER, J. *See* HOFMAN, D.; KRÖGER, F. A.

WAGONER, G. *See* DRESSELHAUS, G.
WALLIS, R. F. *See* BURSTEIN, E.
— (1958) *J. Phys. Chem. Solids* **4** 101: *88, 89, 233*
WALTON, A. K. and MOSS, T. S. (1958) *RAE Rep.* TN/RAD 710: *63*
WANG, C. C. *See* LEVITAS, A.
WARSCHAUER, D. M., PAUL, W. and BROOKS, H. (1955) *Phys. Rev.* **98** 1193: *118, 137*
WARTH, A. H. (1925) *Trans. Amer. Electrochem. Soc.* **47** 62: *102*
WATTERS, R. L. *See* DUNLAP, W. C.; LUDWIG, G. W.
WEDEPOHL, P. T. *See* DYER, H. B.
— (1957) *Proc. phys. Soc., Lond.* **B70** 177: *110, 111*
WEIMER, P. K. (1950) *Phys. Rev.* **79** 171: *158, 160*
— COPE, A. D. (1951) *R.C.A. Rev.* **12** 314: *152, 159, 163*
WEINTRAUB, E. (1913) *J. industr. Engng Chem.* **5** 107: *102*
WEISS, H. *See* WELKER, H.
— (1952) *Z. Phys.* **132** 335: *202, 203, 206*
WELKER, H. and WEISS, H. (1956) *Solid State Physics* **3** 1: Academic Press, N.Y.: *223, 224, 225*
WERTHEIM, G. K. (1956) *Phys. Rev.* **104** 662: *241*
WESTPHAL, W. C. *See* HAYNES, J. R.
WHEATLEY, G. H. *See* HROSTOWSKI, H. J.
WHITE, G. K. *See* FISCHER, G.
WILLARDSON, R. K. *See* HARMAN, T. C.
— HARMAN, T. C. and BEER, A. C. (1954) *Phys. Rev.* **96** 1512: *130*
WILLIAMS, F. E. *See* PIPER, W. W.
— (1957) *J. opt. Soc. Amer.* **47** 869: *207*
WILSON, J. M. (1958) *Progress in Semiconductors* Vol. 3 27; Heywood, London: *113*
WLERICK, G. (1954) *Physica* **20** 1099: *58*
WOLF, M. *See* PRINCE, M. B.
WOLFE, G. A., TOMAN, L., FIELD, N. J. and BRODER, J. D. (1956) *Halbleiter und Phosphore* (Garmisch Semiconductor Conference); Vieweg Braunschweig: *49*
WOLFE, R. *See* AUSTIN, I. G.

WOOD, C. (1956) *Proc. phys. Soc., Lond.* **B69** 613: *199*
WOOD, R. W. (1902) *Phil. Mag.* **3** 607: *158*
— (1933) *Physical Optics*; Macmillan, N.Y.: *10*
WOODBURY, H. H. *See* NEWMAN, R.
— and TYLER, W. W. (1957) *Phys. Rev.* **105** 84: *148*
WOODRUFF, R. W. and GIVENS, M. P. (1955) *Phys. Rev.* **97** 52: *173*
WOODS, J. F. (1957) *Phys. Rev.* **106** 235: *199*
WOODS, S. B. *See* FISCHER, G.
WOODWARD, A. M. *See* MOSS, T. S.
WOODWARD, P. M. *See* BELL, D. G.
WRIGHT, R. W. and BASTIN, J. A. (1958) *Proc. phys. Soc., Lond.* **71** 109: *202*
WURST, E. C. *See* LASSER, M. E.

YAKAMA, E. and SUGITA, T. (1953) *Phys. Rev.* **90** 992: *113*
YAKOBSON, M. A. *See* GROSS, E. F.
YAMANOUCHI, C. *See* FUKUROI, T.
YOSHIDA, S. *See* UNO, R.
YOSHIMATSU, S. *See* IBUKI, S.
YOSHINAGA, H. (1955) *Phys. Rev.* **100** 753: *22, 51, 211, 225, 234*
— and OETJEN, R. A. (1956) *Phys. Rev.* **101** 526: *32, 51, 234, 237*
YOUNG, A. S. *See* LAWSON, W. D.

ZEIGER, H. J. *See* DEXTER, R. N.; LAX, B.
— LAX, B. and DEXTER, R. N. (1957) *Phys. Rev.* **105** 495: *89*
ZITTER, R. N. *See* KURNICK, S. W.
ZWERDLING, S. *See* KEYES, R. J.; ROTH, L. M.
— KEYES, R. J., FONER, S., KOLM, H. H. and LAX, B. (1956) *Phys. Rev.* **104** 1805: *89*
— and LAX, B. (1957) *Phys. Rev.* **106** 51: *90, 133*
— — and ROTH, L. M. (1957) *Phys. Rev.* **108** 1402: *90, 134, 137, 228, 234*
— — — (1958) *Phys. Rev.* **109** 2207: *48, 90, 134, 137*

SUBJECT INDEX

277

CPSIA information can be obtained
at www.ICGtesting.com
Printed in the USA
BVHW051054140223
658483BV00003B/161